T0123003

Early FM Radio

Early FM Radio

Incremental Technology in Twentieth-Century America

GARY L. FROST

The Johns Hopkins University Press
Baltimore

Printed in the United States of America on acid-free paper
2 4 6 8 9 7 5 3 1

The Johns Hopkins University Press
2715 North Charles Street
Baltimore, Maryland 21218-4363
www.press.jhu.edu

Library of Congress Cataloging-in-Publication Data

Frost, Gary Lewis.
Early FM radio : incremental technology in twentieth-century America / Gary L. Frost.
p. cm.
Includes bibliographical references and index.
ISBN-13: 978-0-8018-9440-4 (hardcover : alk. paper)
ISBN-10: 0-8018-9440-9 (hardcover : alk. paper)
1. Radio frequency modulation—Transmitters and transmission—History. I. Title.
TK6547.F76 2010
621.384'152097309041—dc22 2009026947

A catalog record for this book is available from the British Library.

Nor is there any harm in starting new game to invention;
many discoveries have been made by men who were à la chasse
of something very different.

Horace Walpole to Hannah More, 10 September 1789

CONTENTS

ACKNOWLEDGMENTS

The author wishes sincerely to thank the many individuals who kindly assisted him in researching, writing, and critiquing this book. They include Alex Roland, Michael McVaugh, Sy Mauskopf, John Kasson, Peter Filene, William Trimble, Steven Niven, William E. Leuchtenberg, Dana Raymond, John Hepp IV, Stephen Pemberton, Molly Rozum, Larry Wright, Michele Strong, Patrick Sayre, Mary Jane Aldrich-Moodie, David Wunsch, Charles Ritterhouse, Mischa Schwartz, Laura Moore, Anne Langley, Chuck Alley, Robert J. Brugger, Josh Tong, Donna Halper, Harry Maynard, Janice Wright, Linda Schoener, Susan Douglas, Kristen Haring, Brian MacDonald, Houston Stokes, Hans Buhl, David Burke, and James Hoogerwerf, and Richard Boyle.

The Mellon Foundation, the IEEE History Center, and the Society for the History of Technology provided funding for this research project. Karen Maucher, Robert Guy, Marco Greco, and David West helped the author obtain housing while doing research in New York City.

The author also acknowledges the invaluable service provided by Alex Magoun, of the David Sarnoff Library, as well as the librarians at the University of North Carolina–Chapel Hill, Duke University, and North Carolina State University, and the Columbia University Rare Books and Manuscripts Library. Thanks also to the Rare Books and Manuscripts Library for allowing the use of material from the Edwin Howard Armstrong Papers.

AM	amplitude modulation
AP	Edwin Howard Armstrong Papers, Rare Book and Manuscript Library, Columbia University
AT&T	American Telephone and Telegraph Company
BBA	*Broadcasting-Broadcasting Advertising* (magazine)
BTL	Bell Telephone Laboratories
CBS	Columbia Broadcasting System
CGTFS	Compagnie Générale de Télégraphie Sans Fil (Paris)
cps	cycles per second
EM	electromagnetic
FCC	Federal Communications Commission
FM	frequency modulation
FMBI	FM Broadcasters, Incorporated
FRC	Federal Radio Commission
FSK	frequency-shift keying
GE	General Electric Company
IEEE	Institute of Electrical and Electronic Engineers
IF	intermediate frequency
IRE	Institute of Radio Engineers
JHH	John Hays Hammond Jr. (patent assignee)
NBC	National Broadcasting Company
PIRE	*Proceedings of the Institute of Radio Engineers*
RCA	Radio Corporation of America
RCAC	RCA Communications Company, Inc.
RCAM	RCA Manufacturing Company, Inc.
REL	Radio Engineering Laboratories, Inc.
SSB	single-sideband modulation
TGFDT	Telefunken Gesellschaft für Drahtlose Telegraphie mbH (Berlin)
WE	Western Electric Company, Inc.
WEM	Westinghouse Electric & Manufacturing Company

Early FM Radio

What Do We Know about FM Radio?

> It isn't ignorance that causes the trouble in this world; it is the things that folks know that ain't so.
>
> *Edwin Howard Armstrong, quoting Josh Billings, 1944*

This book presents a clean break from the traditional history of frequency-modulation radio. Some readers will open this volume because they already know the canonical story of FM radio's origins, one of the twentieth century's iconic sagas of invention, heroism, and tragedy. Possibly they learned it from Ken Burns's 1992 documentary film, *Empire of the Air*, or from Lawrence Lessing's "definitive" 1956 biography of FM's inventor, *Edwin Howard Armstrong: Man of High Fidelity.*[1] In any event, all those who have written about the history of FM broadcasting tell more or less the same story: In 1933 the U.S. Patent Office issued patents to Armstrong for his system of "wideband" frequency-modulation radio. More than a decade earlier, everyone else had abandoned FM as impractical, but Armstrong's system astonished the world by suppressing static and reproducing sound with far greater fidelity than AM radio did. The Radio Corporation of America tested the Armstrong system and, after concluding that FM threatened its AM radio empire, RCA not only declined to develop frequency modulation but also tried to suppress it. Nevertheless, Armstrong persevered. Spending much of his personal fortune, he built an experimental broadcast station, which led to the Federal Communications Commission (FCC) establishing the first commercial FM broadcast service in 1940. Afterward, as part of a strategy to cripple FM, RCA refused to pay Armstrong royalties for his invention. In 1948 he sued RCA, a move that cost him far more than he could have expected. Finally, in early 1954, as the trial dragged into its sixth year, a despondent, exhausted, and nearly broke Howard Armstrong took his own life.

Although this tale delivers great emotional power, it actually raises questions more important than the ones it answers. No historian has written more than a paragraph or two about the presumably unfruitful FM research that occurred before 1933, nor has anyone explained a glaring contradiction in the attitude of RCA's managers toward new radio technology during the 1930s: why, if they feared FM, did the firm invest so much during the same period in the far more revolutionary technology of television? We have no idea what steps Armstrong took in developing his system, leaving us at the mercy of facile invocations of Armstrong's "genius" to describe how he invented FM. And no historian has analyzed more than cursorily the patents and technical papers of Armstrong, let alone anyone else involved in FM research. To read the canonical history of FM radio is to explore not so much a history as a technological mythology that pits individualism against collectivism, the independent inventor against the malignant corporation, good against evil.

Today, a huge amount of archival material, scarcely examined since it became available nearly twenty years ago, makes possible a challenge to the canonical history. In 1990 the law firm that represented Armstrong donated his files to the Rare Books and Manuscripts Collection of Columbia University. Because Armstrong obtained a copy of every RCA document related to FM radio when he sued that company, these files—consisting of more than five hundred boxes and dozens of reels of microfilm—make up a complete archive of FM radio research before 1940 within the RCA organization. These documents reveal much that conflicts with the canonical history. For example, RCA and other companies did not give up on FM radio before 1933. Also, RCA did not so much fear FM radio during the 1930s as cultivate an indifference based on ignorance about the Armstrong system.

These documents make possible a fresh and much more careful examination of old sources. Recent historians of science and technology will recognize familiar elements in this book. It argues that FM emerged not so much from the mind of a single man but from a decades-long incremental and evolutionary process involving dozens of individuals. Scholars have shown that social-technological systems as complex as FM radio result from far more complicated processes than merely the straightforward application of laws of nature, and in the shaping of FM radio, nature was again only one factor. Because the development of a technology with even revolutionary potential often requires a long period of gestation before gaining momentum, any number of cultural, political, and commercial interests can heatedly contest how natural laws are framed to make new technologies. Except in the narrowest sense, no one can determine the criteria of what constitutes the "best method" among several competing versions of the same complex technology.[2]

What Are AM and FM Radio?

During the twentieth century, two kinds of modulation dominated radio broadcasting: amplitude modulation (AM) and frequency modulation (FM).[3] They both begin with a continuous radio-frequency sine wave called a "carrier"—that is, a wave of constant amplitude and frequency that oscillates above approximately 100,000 cycles per second (cps).[4] Today, the FCC assigns to each licensed station in the United States a precise carrier frequency. An AM station, for example, might transmit its programs on a carrier frequency of, say, 700 kilocycles per second (700,000 cps). The commission regulates FM radio stations similarly but assigns them much higher carrier frequencies, currently between 88.2 and 107.8 megacycles per second.

FM and AM radio also differ in their means of carrying information. An unmodulated AM or FM transmitter conveys silence by radiating only its carrier wave. Modulation occurs when an audio wave—the electrical analogue of a sound wave—alters either the amplitude, frequency, or phase of the carrier. In the case of AM radio, a modulating audio wave causes the carrier to rise and fall in amplitude, thereby creating an "envelope" that replicates the shape of the original audio wave. By contrast, when an audio wave modulates an FM transmitter, the carrier's amplitude does not change; rather, the so-called instantaneous frequency of the carrier wave increases and decreases with the amplitude of the audio wave.[5] Thus, when the amplitude of the audio wave rises to its maximum positive level, the instantaneous frequency of the transmitted wave increases to a maximum limit. Conversely, when an audio wave descends to its most negative point, the transmitted wave decreases its instantaneous frequency to a minimum value. An audio signal with an amplitude between the minimum and maximum values alters the radio-frequency shift proportionally (see figs. 1, 2, and 3).

Finally, FM stations have substantially wider channel widths than typical AM stations do. A channel is the portion of the radio spectrum that any modulated radio signal requires to convey information. The FCC assigns each licensed AM station a carrier-wave frequency, with two 5-kilocycle "sidebands" on each side of the carrier, making up a 10-kilocycle-wide channel. AM station WLW in Cincinnati, for example, broadcasts a 700-kilocycle carrier but uses frequencies from

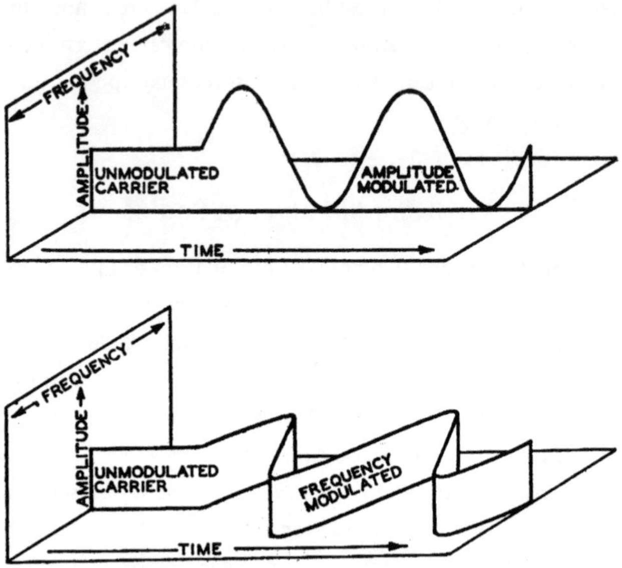

Fig. 1. Comparison of AM and FM Waves. In AM and FM alike, an unmodulated sinusoidal carrier wave radiates at a constant amplitude and frequency. With AM (*top*), modulation occurs when the amplitude of the carrier rises and falls according to the rise and fall of an audio wave. With FM (*bottom*), modulation occurs when the frequency of the carrier "swings" in proportion to the rise and fall of an audio wave. FM sidebands are not shown. Adapted from *The "Radio" Handbook*, 7th ed. (Santa Barbara, Calif.: Editors and Engineers, 1940), 214.

695 to 705 kilocycles to do its job. (To illustrate this arrangement, fig. 4 depicts five channels on the standard AM broadcast band.) By contrast, FM broadcasters use 200-kilocycle-wide channels because those stations effectively emit sidebands that extend 100 kilocycles above and below the carrier frequency.

The Canonical History of FM Radio and Individualist Ideology

Anyone who writes a history of FM radio must come to terms with that technology's canonical history. The narrative of FM radio's genesis, like FM radio itself, evolved from a predicament in which Howard Armstrong found himself during the mid-1930s. In 1934 and 1935 RCA tested his wideband FM system and opted not to purchase the patent rights. The firm never provided clear reasons for spurning FM radio, but this rejection fostered the impression that wideband

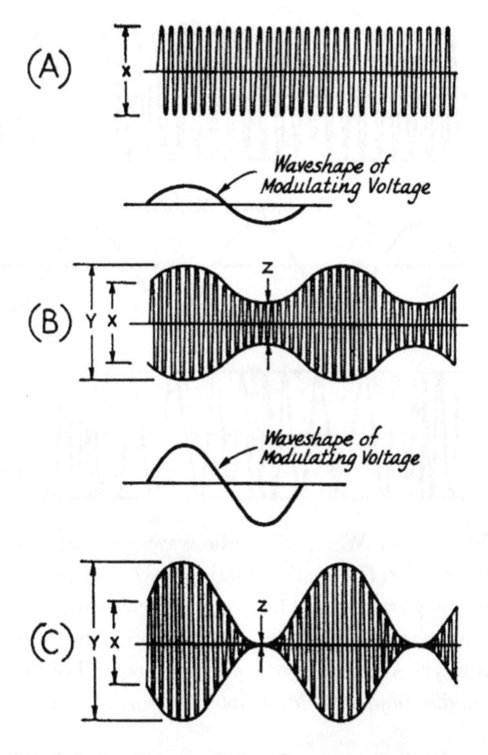

Fig. 2. Amplitude-Modulation Waves. AM radio waves on horizontal time scale: (A) the unmodulated radiofrequency carrier; (B) the carrier at 50 percent modulation; and (C) the carrier at 100 percent modulation. The outline of the modulating voltage is visible on the "envelope" of the modulated carrier waves. Adapted from Headquarters Staff of the American Radio Relay League, *The Radio Amateur's Handbook* (Hartford: American Radio Relay League, 1962), 284.

FM failed on technological grounds, for no other company at the time symbolized more the vibrant technological creativity and expertise that characterized radio. Armstrong and other FM pioneers worked up an alternative explanation in which economic reasons trumped technological ones. Frequency-modulation radio, they claimed, sprang fully developed from the mind of Edwin Howard Armstrong. RCA, which had nothing to do with the origins of Armstrong's invention, declined to back Armstrong out of fear that FM radio threatened RCA's huge capital investment in AM radio technology. For half a century, the boldest and most influential version of this narrative has resided in several chapters of Lawrence Lessing's hagiographic biography of Armstrong. Since then, virtually all historians of FM radio, and consequently anyone who reads about the his-

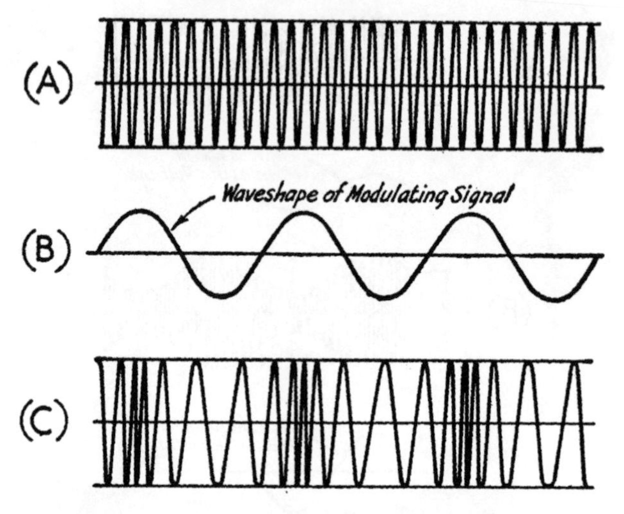

Fig. 3. Frequency-Modulation Waves. FM radio waves on horizontal time scale; (A) an unmodulated carrier; (B) the modulating wave, usually an audio program; and (C) a radiofrequency wave whose wavelength varies with the instantaneous amplitude of B—that is, as B rises and falls in amplitude, the frequency of A rises and falls correspondingly. Adapted from Headquarters Staff of the American Radio Relay League, *The Radio Amateur's Handbook* (Hartford: American Radio Relay League, 1962), 284

tory of frequency-modulation radio, will find himself or herself discoursing with Armstrong, chiefly through the words of Lawrence Lessing.

Lessing spins the invention of FM as a Cold War allegory—an individualist and anticorporate "great man" story, and the culminating episode in the life of a heroic inventor who, "with the pride, secrecy and shrewdness of a lone wolf," patented modern FM in 1933.[6] Armstrong, according to Lessing, represented an earlier period of history that cherished individualism as the cornerstone of American virtue and progress.

> His only faults sprang from his great virtue and strength of purpose. He was a man who would stand up and battle for principles as he saw them against the powers of the world, however formidable. This is becoming so rare a trait as to be prized above rubies. The self-directed individualist, combative, independent and free, who has been responsible for most of the great advances in human culture and invention, is a breed that is passing, at least in this generation and this glacial period of history.
>
> There is, in fact, no one quite of Armstrong's large, individualistic stature left on the inventive scene.[7]

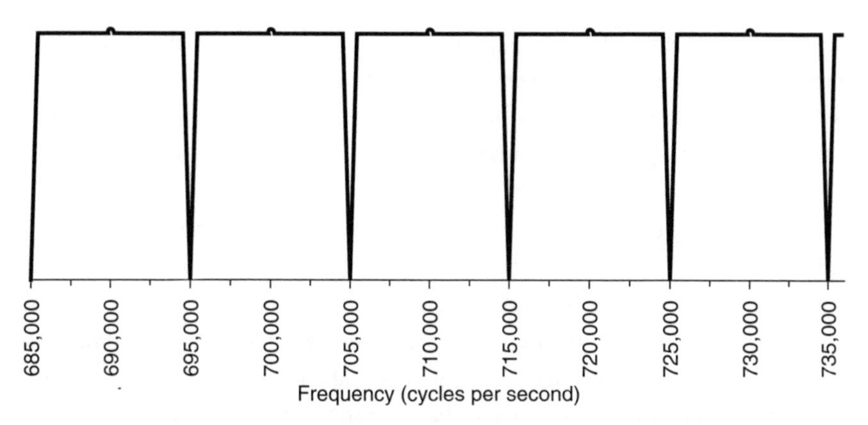

Fig. 4. AM Radio Channels. Depicted in this section of AM broadcast band spectrum are five complete channels and one partial one. Since the 1920s, the Federal Radio Commission and the Federal Communications Commission have allotted the range from approximately 550 to 1,500 kilocycles to broadcasting.

Tragically, he explains, big business interests forced Armstrong to spend the remainder of his life and much of his fortune in court defending his claim to FM as "his last brainchild."[8]

Armstrong's chief antagonist in this saga was the elephantine Radio Corporation of America, ruled "with plump Napoleonic force and immense vanity" by the wily David Sarnoff. RCA occupies a lower level than Armstrong does in Lessing's moral universe, largely because corporations are at best purveyors, not creators, of technological creativity. "It is only the stray, non-conforming individual, rubbing by chance and inclination against freely available knowledge who makes the great discoveries or inventions," he says. "Neither big research teams nor giant laboratories nor large research budgets can substitute for one creative mind. Every great product or development of modern industry may be traced to such independent individuals. Only rarely have they been found in the employ of industrial laboratories."[9]

Lessing reveals almost nothing about FM research before 1933, which seems to buttress his implication that "big research teams" and "giant laboratories" had nothing to do with the development of FM radio. Moreover, he describes RCA and David Sarnoff after that year as obstructing the development of frequency modulation. Fearing that FM might destroy RCA's AM-based empire, Sarnoff betrayed Armstrong—and by extension America—first by withholding RCA's financial backing for FM, then by attempting to "talk down" FM, and finally by trying to rob Armstrong of his claim to FM's invention. In 1948 Armstrong sued

RCA over the firm's refusal to pay royalties for using his FM patents. Delaying tactics on the part of RCA's lawyers prolonged the litigation for six years, which led to tragedy. On the night of 31 January 1954, Howard Armstrong—"at the end of his rope"—stepped to his death from the window of his thirteenth-floor New York City apartment.[10]

Regrettably, the authority accorded to *Man of High Fidelity*, which was published two years after Armstrong's suicide, speaks more to the absence of research since 1956 than to the quality and scope of Lessing's scholarship. To be fair, Lessing frames a nuanced argument that incorporates plausible contextualist themes. Much of his book situates Armstrong amid social and economic forces that few historians today would deny, such as the dramatic growth of corporate research and economic power in the early twentieth century. But Lessing's book, engagingly written for a general readership, discusses sources only sporadically and even admits to bias. The foreword to the first edition describes *Man of High Fidelity* not as a history of FM but rather a "partisan [biography] with respect to the man, whom the author as a journalist knew over a period of fifteen years and esteemed as a great man."[11]

Despite these shortcomings, Lessing's influence on subsequent histories of FM cannot be overstated. Only one historian has used other, mostly primary, sources, to refute a small part of Lessing's version.[12] More typically, the only book-length history of FM published before 2008 cites *Man of High Fidelity* more than any other source, and virtually every radio history Web site recommends the book to visitors curious about FM's origins.[13] Lessing's interpretation has seeped into even the most distinguished scholarship. Thomas Hughes's *American Genesis*, Tom Lewis's *Empire of the Air*, and Susan Douglas's *Listening In* all depend heavily on Lessing.[14] Even Christopher Sterling and Michael Keith, whose recently published book, *Sounds of Change*, constitutes the best general history of FM radio broadcasting, stay close to Lessing's interpretation when examining the prewar period.[15] Indeed, with little else written about the subject, how could they not?

Methodology

This book follows an approach that borrows from scholarship of the past three decades. For several years historians of technology have been classified by how much or little they choose to emphasize the material aspects of their subject. At one end of the spectrum are "internalists" who, as John Staudenmaier writes, "converse with a narrowly defined group of scholars who have made the technology in question their primary concern." Somewhat derisively and unfairly called "gear fondlers"—or "tube fondlers," in the case of radio history buffs—internalists tend to focus almost exclusively and often meticulously on the "nuts and bolts" of their subject. The polar opposites of this group are "externalists," who are "interested in cultural context, [and] pay almost no attention to issues of technological design." Because I find great value in both approaches, this study adopts a middle-ground "contextualist" approach, which, as Staudenmaier explains, "attempt[s] to integrate a technology's design characteristics with the complexities of its historical ambience."[16]

This study also takes what has variously been described as a moderate social-constructivist or socially shaped perspective. Social constructivists renounce historical interpretations based on technological determinism as simplistic. That is, they see technology not only as an explanation for history but also, and more often, as something to be *explained by* historical forces.[17] Accordingly, this book argues that cultural, organizational, economic, and other contingent "social" factors strongly shaped the design of broadcast FM radio at every step. To be sure, technology, including radio, has exerted tremendous influences on society. But social constructivist histories look beyond a one-dimensional "technology-drives-history" perspective, seeking the social factors that influence historical actors to make choices that lead to the development of certain technologies. Therefore, this study takes, in the words of Thomas Hughes, a more or less "seamless web" approach toward its subject, in which the hardware of the technology is both cause and effect in the historical narrative.[18] One can neither remove the social from the technological side of FM's history nor remove the technological from the social.

This is not to say that FM was entirely socially constructed, with the natural world having no say in the matter, but only that no one "discovered" FM in the same way that William Herschel discovered the planet Uranus or Glenn Seaborg discovered plutonium. This study tells not so much a history of the "construction" of a technology as a "shaping" of a technology, because readers might infer from the word "construction" that the natural world plays no role in how any technology is interpreted.[19] On the contrary, making complex new technological systems resembles collective artistic creativity more than scientific discovery. Each member of a group of sculptors can take a turn at chiseling a block of marble to "reveal" the statue underneath. Every artist—every group of artists—will "find" a different figure but must do so within two constraints, one contingent and flexible—human imagination—and the other mercilessly rigid, namely, the marble's natural properties. Invention also must comply with laws of nature, some of which are imperfectly understood, or even unknown, but those laws rarely restrict the process of developing new technology sufficiently to allow anyone to predetermine the outcome. Thus, FM has come in innumerable forms, only one of which makes up modern broadcast FM, which itself differs significantly from what Armstrong's 1933 patents described. In other words, the invention and development of FM followed a trajectory constrained by the natural limits of the material world, planned research, and happenstance.

Finally, this study will delve into the technical details of radio. The economic historian Nathan Rosenberg has written that "the social and economic history of technology can only be properly written by people possessing a close familiarity with the actual technology itself."[20] Fortunately, this book demands far less from the reader, who will, with patience, nevertheless learn a little about how radio works. Novices to the field should take comfort in the fact that much of the technology itself was literally child's play. After the crystal detector was patented in 1906, the ranks of amateur radio operators swelled with hundreds of thousands of boys (and more than a few girls).[21] The apparatus they "worked" was largely composed of simple hardware: wires, condensers, transformers, and insulating materials, for example. Assembling these components into a radio set was a creative process, but not an illogical or an especially complex one. True, a schematic drawing of an early radio transmitter or receiver could baffle anyone entirely untrained in circuit theory, but with a little effort, young practitioners learned that becoming an expert chiefly required comprehending a few rules about a few electrical components. Still, this study does not assume that its readers have reached the same level of technical expertise as the child hobbyists of a century past and will from time to time translate "texts" of early twentieth-century radio technology that appeared in patents, published technical papers, and personal correspondence.

A Note about Terminology

Decades of imprecise terminology have muddled FM radio's history. Traditionally, the system for which Armstrong was awarded patents in December 1933, and on which modern broadcast FM has been based since 1940, has been called both *Armstrong FM*, and *wideband FM*. The usage of wideband FM has fostered the inference that previous, presumably failed, FM systems were narrowband FM, a term that suggests a channel width narrower than a standard 10,000 cps AM band. I am forced to employ the term wideband FM, but I also emphasize that only a negligible proportion of FM research ever targeted a channel narrower than 10,000 cps.

TABLE 1
Categories of FM Radiotelephony

Ehret FM (1902)
Existed only in two patents issued to Cornelius Ehret. Spark gap transmitter with slope detector receiver. Unworkable.
Narrowband FM (early to mid-1920s)
Channel width: 10,000 cps or less. Determined to be theoretically unworkable in 1922, although the U.S. Patent Office later issued four narrowband FM patents.
Armstrong wideband FM (patents filed January 1933)
Armstrong low-tube-hiss FM (January 1933–Spring 1934)
Described in Armstrong patents of 26 December 1933 as a system for the reduction of tube hiss. Frequency swing: 150,000 cps. Armstrong originally declared that his invention had no effect on static.
Armstrong low-static FM (Spring 1934–)
Essentially the same technology as low-tube-hiss FM but reinterpreted. During the spring of 1934, Armstrong learned that, contrary to what he had claimed in his patents, his invention reduced static dramatically.
Armstrong high-fidelity FM (late 1937–)
Low-static FM after Armstrong began incorporating high-fidelity audio circuits into his low-static system. Audio bandwidth was an unprecedented 15,000 cps. Functionally equivalent to modern monophonic broadcast FM radio.
All other frequency-modulation and phase-modulation systems (1920s–)
All other FMs, both before and after the invention of Armstrong wideband FM. Channel widths ranged between 10 and 30 kilocycles.

Moreover, the terms Armstrong FM and wideband FM falsely imply a nonexistent stability for the technology. In fact, Armstrong's understanding of what he had made changed significantly during the period from 1933, the year he filed his most important FM patents, and 1940, when the FCC established a new FM broadcast radio service. For that reason, I use Armstrong FM (and wideband FM) only as general terms for the invention for which he was awarded patents in December 1933 and which continued to evolve afterward. But I also use *low-tube-hiss FM*, *low-static FM*, and *high-fidelity FM* to refer to Armstrong FM at three stages of its development during the mid- and late 1930s (table 1).

Finally, the reader should note that this book conforms to a convention of the period it covers by not using the modern unit of frequency, the Hertz. Instead, frequency is measured, as it was before 1960, with one of the following units: cycles per second (cps) or cycles; kilocycles per second (kps) or kilocycles; or megacycles per second or megacycles (mc).

CHAPTER ONE

AM and FM Radio before 1920

The process of altering the length of the emitted wave must be abandoned.

Valdemar Poulsen, inventor of FM radiotelegraphy and its first critic, 1906

The Spark Gap and the Coherer

To understand why frequency-modulation radio first appeared in 1902, one must know something about the technological context of the radio at that time. Two devices—the spark gap, used in transmitters, and the coherer, the basis of almost all early wireless receivers (figs. 5 and 6)—had defined the possibilities and the limitations of the art since Guglielmo Marconi invented radio during the 1890s.[1] The simplest form of spark gap featured a pair of spherical brass electrodes separated by one or two inches of air. When a battery-and-coil circuit caused the electric potential (i.e., the voltage) between the electrodes to rise above a certain threshold, a spark leaped across the gap, discharging violently and emitting a train of *damped waves*—invisible electromagnetic (EM) waves that decrescendoed to nothing in a fraction of a second (fig. 7). The phenomenon resembled the dropping of a stone into a still pool of water, or a clapper striking a bell.

To send the dots and dashes of Morse code messages, Marconi borrowed a method of modulation from overland wire telegraphy. For decades telegraph operators had utilized a key—a hand-operated electrical switch—to signal either full-power *marks* when current was on or zero-power *spaces* when current was off. Though unnamed, this method could be described as a type of binary amplitude modulation—that is, transmission occurred either at full amplitude or

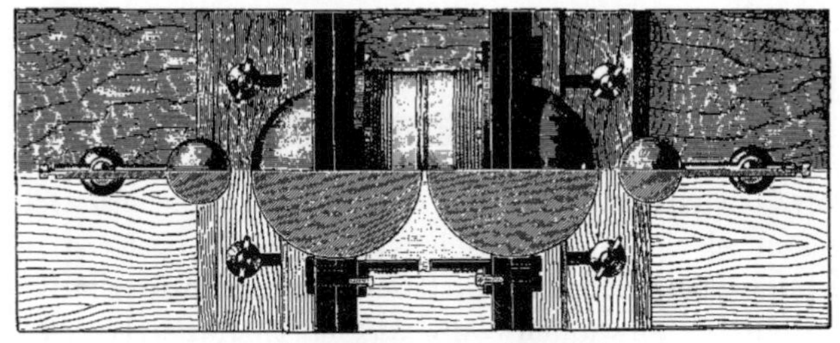

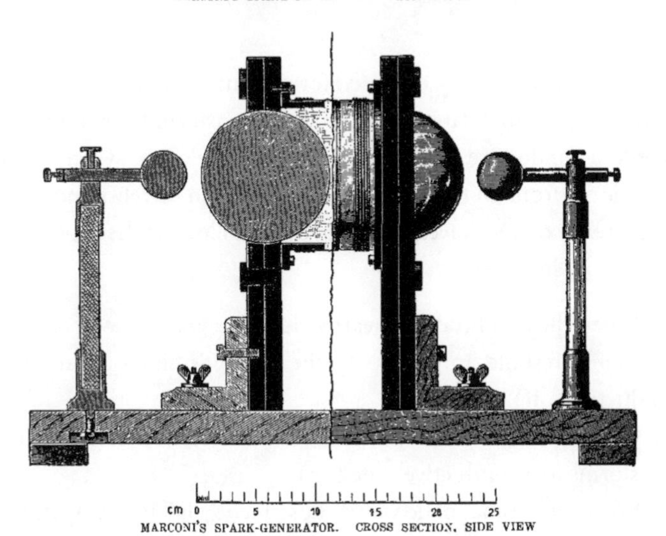

Fig. 5. Marconi Spark Gap Transmitter, 1898. Depicted are two gaps between metal spheres. Other versions of the device had one or several gaps. Adolph Slaby, "The New Telegraphy," *Century* 55 (April 1898): 880.

not at all, a short mark corresponding to a dot, a somewhat longer mark a dash. Binary amplitude modulation required substantial modifications before Marconi could adapt the method to wireless. A single train of damped waves sufficed to signal a dot, but several trains had to signify a dash, which necessitated rapid refiring on the part of the spark gap. Marconi therefore designed his transmitter to recharge the spark gap quickly and automatically, so that holding the key down caused the transmitter to sputter out continuous trains of damped waves, one closely ranked group after another.

In the receiver side of his system, he complemented the spark gap with a coherer. This was a hollow glass tube, approximately six inches long and packed with metal filings (Marconi preferred a mixture of nickel and silver). Metal plugs at each end compressed the filings and functioned as electrical terminals besides.

Fig. 6. Branley Coherer, 1902. Inside an evacuated glass tube are metal filings. Silver plugs act as terminals and enclose the filings by capping the open ends of the glass tube. Detail from figure in Ray Stannard Baker, "Marconi's Achievement: Telegraphing across the Ocean without Wires," *McClure's Magazine*, February 1902, 291.

If all went well—which rarely occurred for reasons discussed later—the coherer detected radio waves that caused the filings to transform from a nonconductive (off) state to a conductive one (on), much like an electric switch. An opposite transformation—from on to off—required more than merely removing the tube from the presence of EM waves, however, as the following description of an off-on-off sequence illustrates:

Step 1. When the amplitude of nearby electromagnetic waves remains below a certain threshold amplitude A_n, the coherer's filings are normally nonconductive (off).

Step 2. When electromagnetic waves rise above A_n, the coherer instantaneously transforms to a conductive state (on).

Step 3. When electromagnetic waves subsequently fall below A_n in amplitude, the coherer remains on—until some physical motion disturbs the arrangement of the metal filings.

Step 4. Tapping the side of the coherer, therefore, causes the device to revert to a nonconductive state (off).

Step 5. The newly nonconductive coherer now exists in the same nonconductive state as in step 1, but with a new threshold amplitude, A_{n+1}. Its value corresponds to the arbitrary physical rearrangement of its filings and might differ substantially from A_n.

It should be emphasized that the coherer could not detect messages by itself because, once the device turned on, it stayed in that state until two events took place. First, local electromagnetic waves had to subside in strength below the trigger threshold A_n; and, second, some material object had to jar the internal metal filings with force adequate to alter their physical arrangement, causing the coherer to revert to a state of nonconductivity. To meet the second requirement, Marconi fastened to the coherer a "vibrator," or "tikker"—a small electrically driven hammer mechanism that continuously rapped the outside of the glass

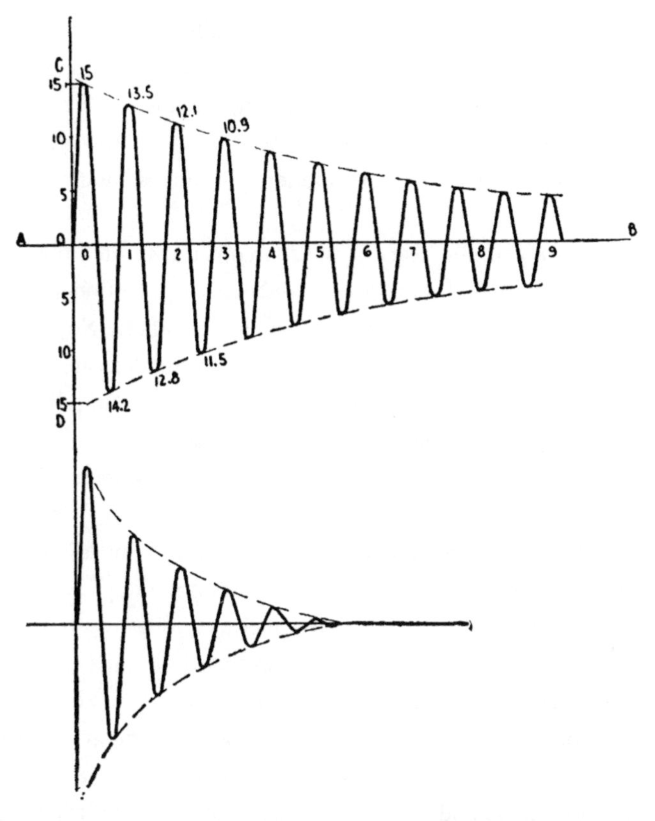

Fig. 7. Damped Waves. Two trains of waves illustrate "feebly damped" (*top*) and "strongly damped" (*bottom*) waves. Dotted lines trace the "envelope" of the wave train peaks. Elmer E. Bucher, *Practical Wireless Telegraphy: A Complete Text Book for Students of Radio Communications* (New York: Wireless Press, 1921), 1.

tube. Finally, to record messages, Marconi wired his coherer-switch to operate either a buzzer or a writing device, usually a paper-tape inker.

This was the theory of Marconian wireless telegraphy, but in practice the spark gap and coherer amounted to what Thomas Hughes has termed "reverse salients" and Edward Constant calls "presumptive anomalies."[2] That is, practitioners by and large knew that no matter how much the two devices could improve, no one would ever overcome their inherent limitations, and sustain progress in the art. The coherer's fickle responsiveness especially rankled operators. Each tap of the vibrator caused the filings to jump from one level of sensitivity to another with kaleidoscopic capriciousness. At one moment, for example, a coherer might fail to detect a transmitter situated only yards away, but a fraction of a second later electromagnetic waves from a source dozens of miles distant could trigger the filings into a conductive state. Further, the coherer, as a two-state binary switch, proved entirely inadequate for wireless telephony, because replication of speech required a continuously and proportionally responsive detector—that is, a detector capable of tracking variations corresponding to the instantaneous amplitude of sound.

As for the spark gap, the heart of the Marconi transmitter, the amplitudes of its damped waves *could* be varied—modulated, as practitioners say—which tempted some to try adapting the device to wireless telephony. No one succeeded entirely, but the Canadian-born inventor Reginald Fessenden came close. In December 1900 he transmitted speech on Cobb Island, Maryland, by amplitude-modulating a continuously triggered spark gap. Fessenden obtained only minimally intelligible reception, though, on account of static-like noise created by the spark.[3]

Nor could spark gaps and coherers overcome natural impediments to radio wave propagation. Electromagnetic radiation does not always move along predictable paths. Local weather, upper-atmospheric conditions, sunlight, and the lengths and amplitudes of the waves themselves chaotically affect the attenuation and refraction of electromagnetic waves. A large mass, such as a hill, office building, or forest, for example, can absorb electromagnetic energy or create a virtual mirror that causes waves to carom off in another direction. Although early wireless pioneers understood the laws of propagation poorly, if at all, experience quickly taught them two dismal facts: first, operating more than one transmitter on the same wavelength invited interstation interference; and, second, the strength of a signal at the receiver might fade, which multiplied the ill effects of the coherer's notoriously erratic sensitivity.

Tuning and the Resonant (*LC*) Circuit

Resonance, the principle behind tuning, also figured prominently in the technological context of early radio. All circuits, whether a piece of wire or a complete radio transmitter, possess the complementary reactive properties of inductance, symbolized by *L*, and capacitance, symbolized by *C*. (Circuits also contain resistance, which we neglect here for the sake of simplicity.) Stimulating a circuit electrically causes it to resonate naturally at a specific wavelength. In electrical circuits, this mathematical formula defines the length of the resonant wave

$$\lambda_i = \frac{c}{f_i} = 2\pi c \sqrt{L_i C_i}$$

where:

λ_i = instantaneous resonant wavelength (measured in meters),
f_i = instantaneous resonant frequency (measured in cps),
L_i = instantaneous circuit inductance (measured in henrys),
C_i = instantaneous circuit capacitance (measured in farads),
π = a constant, approximately 3.14, and
c = the speed of light, a constant equal to 299,800,000 meters per second.

This formula boils down to some simple concepts. Because c and π are constants, the wavelength λ_i is proportional to $\sqrt{L_i C_i}$. Thus, increasing either L_i or C_i will increase λ_i (and decrease the resonant frequency, f_i). Conversely, if one decreases either L_i or C_i, λ_i will also decrease.

Traditionally, practitioners have used the preceding resonance formula far more often to understand resonant circuits than to construct them. As the historian of radio Sungook Hong observes, tuning was primarily "not a mathematical principle but a *craft*."[4] Early wireless pioneers devoted much of their time to perfecting practical techniques for adding reactive components—either inductive or capacitive—to a circuit for the purpose of adjusting it to resonate at a precise wavelength. Amateur radio operators usually accomplished this by making their own components. Coiling wire around a cylindrical oatmeal box or an iron core, for instance, provided inductance. A stack of metal plates, each sandwiched between layers of air, oil, or paper, made up a condenser, the component that supplied capacitance. Tuning to a particular station was commonly accomplished by making the inductive element of an *LC* circuit variable and adjusting the inductance *L* until the circuit resonated at the wavelength of the station's frequency. One could do the same thing with the capacitive element by mounting on an axle half of several interleaving plates constituting a condenser. Rotating the axle changed the capacitance *C* and thus the wavelength. Until a few years ago all consumer radio receivers employed fixed-value inductance and a mechanically variable condenser for tuning purposes. A knob on the front panel of a home radio set was fastened to the axle of a variable condenser. Listeners tuned their radios to a station by adjusting this knob to a number on a dial that corresponded to the station's carrier frequency—a method still employed in cheaper radio receivers. Today, electronic circuits have replaced this mechanical arrangement to do the same thing.

Cornelius Ehret and the Invention of Frequency-Modulation Radio

Although most histories of frequency-modulation radio state that Armstrong invented the technology in December 1933, FM appears more than thirty years earlier in American and Danish patent records. On 10 February 1902 a Philadelphian named Cornelius Ehret filed a patent application for a frequency-modulation system.[5] Seven months later, Valdemar Poulsen applied for a Danish patent for a radio-frequency "arc oscillator" that also employed FM.[6] These two men shared little beyond being contemporaries, however. Ehret began as an unknown amateur and, despite his invention, remained so, but Poulsen had already achieved international recognition by inventing magnetic recording. Further, Ehret explicitly claimed to have invented a frequency-modulation system of wireless telegraphy and telephony, though he failed to make a functional prototype. By contrast, Poulsen eventually made a radiotelegraph system using a method of frequency modulation that he would renounce, but which practitioners would copy for more than two decades.

As the first American holder of a frequency-modulation radio patent, Ehret ranks among the most obscure inventors in the history of wireless. Except for a short article that appeared nearly seventy years ago in *Communications* magazine, virtually no twentieth-century history of radio mentioned him.[7] He does appear in patent court records, however. In 1959, New York's Southern District Judge Edmund L. Palmieri decided in favor of Armstrong's patent infringement suit against Emerson Radio, which had cited Ehret's system to dispute the novelty of wideband FM. Palmieri acknowledged that "the Ehret patent was one of the earliest patents in which it was proposed to transmit and receive intelligence by varying the frequency of a radio wave," but he dismissed outright Ehret's influence on modern FM radio by declaring that "the Ehret patent did not teach anything at all concerning the problem of reducing static and noise in radio signaling. It did not refer to, or suggest anything concerning, the bandwidth to be employed in frequency modulation or the extent of variations in frequency to be employed. It did not refer to or suggest limiting in a frequency modulation receiver." In other words, the fact that Ehret neglected to specify a channel width, employ a limiter circuit, and claim a reduction in "static and noise"—all features that Judge Palmieri attributed to Armstrong FM— banished Ehret to the backwaters of history. "There is no evidence," Palmieri concluded, "that the Ehret patent had any impact upon the art."[8]

The conclusion is fair enough insofar as the law goes, and this study does not challenge Palmieri's assertion that Ehret left no impression on the art of radio design. But historians should not leave the evaluation of Ehret's historical significance to lawyers. Patent courts make winner-take-all decisions chiefly by weighing competing claims of priority, and whether an invention "works," criteria that ignore crucial questions of historical interest that go beyond, say, which individual should be given all the credit for inventing a particular technology. For what purpose, for example, did Ehret envisage his invention? What did his invention reveal about the state of the art of wireless—what some would call the "culture" of wireless—during the first decade of the twentieth century? And what was the relationship, if any, between his FM and the kinds of FMs that followed? Did Ehret discover anything inevitable about frequency modulation?

Because Ehret's patents constitute the entire record of his career, answering these questions presents difficulties. Nevertheless, as the historian Eugene Ferguson has similarly demonstrated for mechanical engineering drawings and architectural plans, Ehret's patents, when carefully decoded in context, reveal far more information than the fact that their inventor made an impractical invention that had little effect on later work.[9] They illuminate, for example, the tacit knowledge of wireless engineering during a period that paved the way for modern FM radio. Virtually all wireless pioneers, and most electricians in 1902, would have understood implicitly the symbolic language of Ehret's circuit schematics, and why he connected spark gaps, condensers, hand-wound inductors and transformers, "air-gaps," tuned circuits, wires, and "telephone-receivers" in the ways he did.

That Ehret invented not only frequency-modulation telegraphy but also a radiotelephony system is indisputable. On 28 March 1905 the U.S. Patent Office, which had divided his original application, issued to the Philadelphian a pair of almost identically worded patents for a system that transmitted and received "the reproduction of speech and other signals through the agency of means responsive to changes or variations in the frequency of the received energy."[10] Although Ehret never explicitly articulated the motivations behind his invention, clearly he sought to overcome two difficulties associated with wireless. One was fading, which still plagues electromagnetic communications. Discarding the skittish coherer, whose electrical properties transformed with every tap of the tikker, Ehret combated fading by designing instead a wireless telegraph receiver with rock-steady sensitivity (figs. 8 and 9). Moreover, unlike the coherer, which detected only the presence and absence of waves, his receiver contained a resonant *LC* circuit that attenuated incoming EM waves roughly in proportion to their length. For instance, suppose capacitance 39 and inductance 40 in figure 8 are chosen so that the circuit resonates with waves 340 meters in length. A wave with

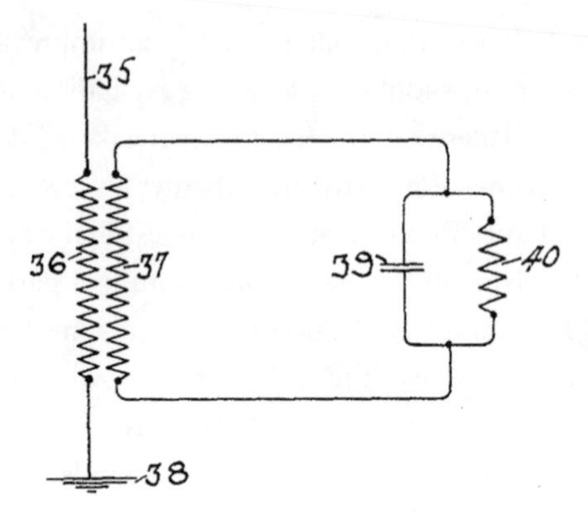

Fig. 8. Detail from Ehret Slope Detector, 1902. In this Ehret frequency-modulation receiver, 36, 37, and 40 specify inductors, and 39 is a condenser. Cornelius D. Ehret, "Art of Transmitting Intelligence," U.S. Patent No. 785,803, application date: 10 February 1902, issue date: 28 March 1905.

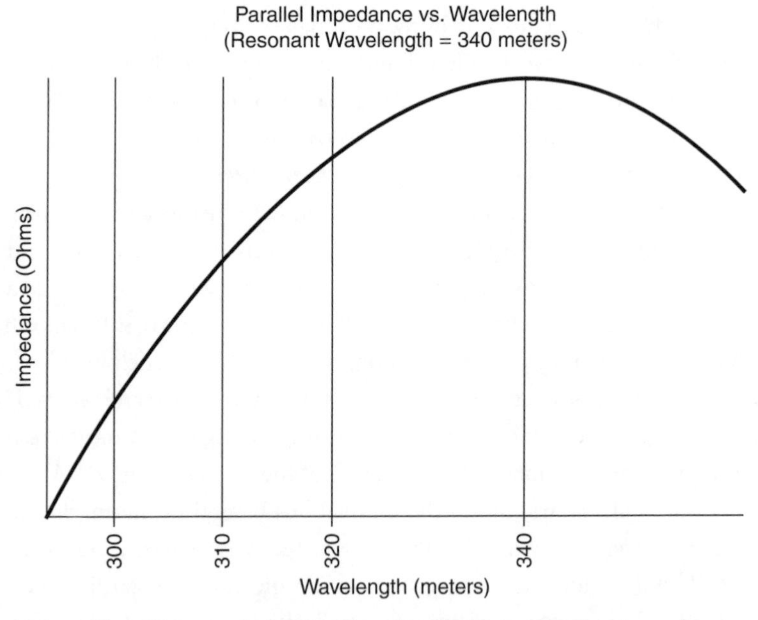

Fig. 9. Ehret Slope Detector Response, 1902. In this depiction of an incoming wave amplitude across a parallel *LC* circuit modeled on Ehret's frequency-modulation slope detector, the circuit is tuned to a resonant wavelength of 340 meters, at which the relative amplitude is maximum. Amplitude decreases approximately linearly from 320 to 300 meters. For telegraphy, a transmitter radiated 300-meter waves to indicate a mark; 310-meter waves indicated a space. For telephony, the instantaneous audio amplitude is proportional to the instantaneous positive-or-negative deviation from a center (i.e., the reference) wavelength of 310 meters.

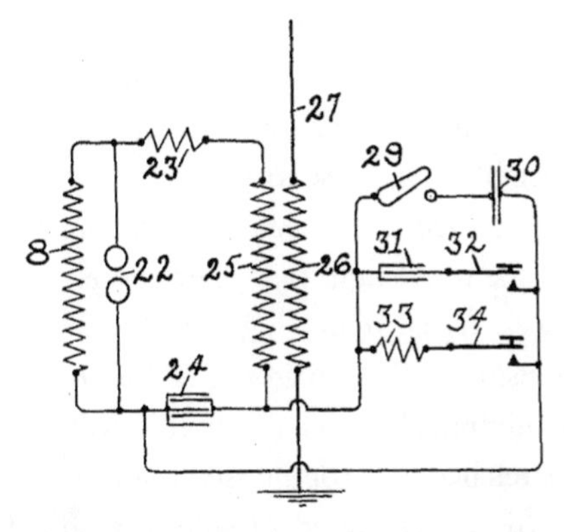

Fig. 10. Detail from Ehret Transmitter, 1902. Drawing of an Ehret frequency-modulation transmitter illustrates two methods of telegraphy and one method of telephony: 22 is a continuously triggered spark gap (trigger mechanism not shown); inductors 8, 23, and 25 constitute the nominal circuit inductance; 24 is the nominal circuit capacitance, which together with the inductance causes the circuit to radiate a wave with a fixed length; and 32 and 34 are telegraph keys that switch condenser 31 and inductor 33 respectively in and out of the circuit, thereby shifting the wavelength of the circuit slightly. This is FSK modulation. For radiotelephony, element 29 switches telephone transmitter (i.e., the microphone) 30 in and out of the circuit. If the microphone is of the inductance type, speaking into it causes the overall inductance of the tuned circuit to alter with the instantaneous amplitude. Thus, the instantaneous frequency of the tuned circuit varies with the amplitude of the speech. Cornelius D. Ehret, "Art of Transmitting Intelligence," U.S. Patent No. 785,803, application date: 10 February 1902, issue date: 28 March 1905.

a wavelength of 340 meters is minimally attenuated, and therefore its amplitude as measured across 39 and 40 is maximum. If the wave decreases in length, say to 300 meters, the circuit will attenuate the wave more, causing the amplitude across 39 and 40 to decrease as well.

Ehret's second goal was to design a system that transmitted and received wireless telegraph and telephone messages. In doing so, he borrowed from Marconian technology, even as he tried to overcome its limitations. Although his telegraph retained the spark gap, his transmitter (fig. 10) radiated damped waves nonstop, as opposed to Marconi's system, which sparked only during keying. Further, Ehret's telegraph radiated waves with either of two lengths, in contrast to the single-wavelength trains of a Marconi transmitter. The patents neglected to mention specific values of capacitance and inductance, so we cannot determine even the range of wavelengths Ehret had in mind, but this time, again for the sake of illustration, let us assume that 300 meters represented a space and

320 meters a mark. To achieve this, the aggregate effects of inductors 8, 23, 25, and 26, and condenser 24 constituted a resonant circuit that radiated 300-meter waves. Switching another inductor or condenser into the circuit caused the circuit's resonant wavelength to increase to 320 meters. Ehret employed telegraph key switches to do this: switch 32 connected and disconnected the condenser 31, and switch 34 caused the wavelength to toggle between 300 meters (space) and 320 meters (mark) by electrically removing and adding inductance 33.

The telephone transmitter resembled the telegraph but with an important difference. Instead of jumping between two fixed wavelengths, Ehret's transmitter instantaneously stretched and compressed the length of its outgoing wave to correspond proportionally to rapid variations of sound. Opening switches 32 and 34 (fig. 10) removed capacitance and inductance used only for telegraphy. Then, closing switch 29 connected to the circuit element 30, a microphone whose diaphragm deflected in proportion to the instantaneous amplitude of the sound. Because the diaphragm was mechanically coupled to a variable inductor (or variable condenser, depending on a designer's preference), the *L* (or *C*) of a tuned *LC* circuit was proportional to the sound's instantaneous amplitude. As an illustration, if the microphone detected no sound, the transmitter radiated waves with a length of, say, 310 meters, the nominal length for the transmitter's tuned circuit. At maximum amplitude, the microphone's inductance wobbled between two extremes and thus pushed and pulled the wavelength to maximum and minimum values of, say, 300 and 320 meters. Similarly, a midlevel amplitude would cause the wavelength to wobble between 305 and 315 meters.

Although no evidence exists that anyone knowingly copied his circuits, Ehret anticipated much that appeared in frequency-modulation systems of several later decades. His idea to link mechanically a microphone to a condenser or inductor would be replicated in most FM radiotelephony patents through the mid-1920s. And even after electronic amplification revolutionized radio circuit design after 1920, FM detectors that resembled Ehret's receiver—called "slope detectors" by then—appeared in systems well into the 1940s.[11] More impressively, Ehret was the first to modulate the length of a transmitted wave by altering the inductance or capacitance in a resonant *LC* circuit, a practice that survives today in radiotelegraphy as "frequency-shift keying" (FSK). Because marks and spaces correspond to different wavelengths, a receiving station operator can distinguish a transmitting station that has gone off the air from one that has simply halted transmission temporarily. A receiving station operator who detects a space wave that lasts several minutes can be certain that the sending station's signal has not faded away and that the transmitter's operator has stopped keying. Ehret never pointed out this advantage, though, possibly because he never realized it.

Despite its novelty, the Ehret system exemplifies how even the seemingly most innovative technological innovations draw primarily on traditional ideas. Marconi's first wireless telegraph—an invention that wrought radical changes on the world if one ever did—borrowed liberally from the decades-old practices of electrical engineering and overland telegraphy. The very ordinariness of the Ehret patents also shows how inventors lean far more toward the evolutionary than the revolutionary. Ehret worked well within the normal practice of electrical technology, using the already-venerable resonant circuit, a device that will probably continue to survive for several decades, if not centuries. Anyone familiar with the visual language of electrical engineering in 1902 would have found no basic device in the Ehret patents that had not been previously used elsewhere. Moreover, the staying power of Ehret's circuits also confirms how an innate conservatism characterizes technological innovation. Not until at least the 1950s did FM Ehret's slope detector and reactance microphone fall out of normal practice.

Ehret's FM also exemplifies the fact that historical and technological contexts shape how problems that technologies purport to solve can wax and wane in importance. What seems an urgent issue at one point can recede in significance a short time later, thus causing innovators to abandon technological paths that they had previously hoped would lead to a solution. Or, perhaps, other technologies alleviate the same problem more effectively. In 1902, when wireless communication rarely extended beyond a few miles, Ehret undertook to defeat with frequency modulation the exasperating tendency of radio signals to fade in strength. Conceptually he was on the right track. Modern FM compared with AM resists fading extraordinarily well because of its inherent insensitivity to amplitude variations, and perhaps someone might have used Ehret's ideas as a stepping-stone toward a practical system of frequency modulation. But only a few years after those patents were filed, other technological improvements, such as electronic amplification and directional antennas, made for stronger signals and mitigated fading sufficiently to cause the problem to decline in importance.

Ironically, one feature that made Ehret's inventions exceptional also accounts for their fundamental impracticability. Ehret used an *LC* circuit for a receiver because, unlike the coherer, a resonant circuit responds to changes in the wavelength of the incoming wave. But *LC* circuits detect amplitude fluctuations as well as frequency swings. In other words, his slope detector was both frequency- *and* amplitude-responsive, making it as vulnerable to fading as any other detector.

Modern practitioners will find it difficult to imagine how Ehret could have overlooked this flaw, and soon after electronic vacuum tubes became widely available during the 1920s, FM researchers found a solution. They compensated for amplitude fluctuations with an electronic circuit that automatically raised and lowered the amplitude of incoming radio waves to a fixed voltage—the very same "limiter" that Judge Palmieri in 1959 found wanting in Ehret's patents and which Palmieri mistakenly implied that Armstrong had invented for the first time in 1933.[12]

Valdemar Poulsen and Frequency-Modulation Radiotelegraphy

Historians of radio have recognized the other earliest inventor of FM, the Danish engineer Valdemar Poulsen, for his pioneering work with the arc oscillator, one of the most important devices of the early wireless era, but almost no one has mentioned that Poulsen incorporated frequency modulation into his invention.[13] The arc earned its prominence because it emitted relatively low-distortion continuous-wave radio frequencies at previously unattainable levels of wattage. An ideal continuous wave is perfectly sinusoidal, and by 1902 engineers had essentially met that standard with electromechanical alternators that delivered 50 and 60 cps electrical power. But wireless communications required at least a thousand times those frequencies, which presented the daunting problem of making an alternator spin fast enough without flying to pieces. Eventually the General Electric Company (GE) manufactured high-frequency alternators that achieved 200,000 cps.[14] But until the wide use of electronic vacuum tube oscillators during the twenties, only the arc created close-to-sinusoidal radio waves in frequency ranges above 500,000 cycles per second.

Poulsen borrowed both the arc and frequency modulation from the field of electrical music, which itself descended from the arc light of the nineteenth century. First developed by the English physicist Humphry Davy about 1808, arc lights illuminated vast areas by forcing a large continuous electrical current to flow across a gap of air separating two carbon electrodes. (The current appeared to follow a curved path, which accounted for the device's name.) The arc was characterized by not only its blinding brilliance but also an audible hiss, which, as the English physicist William Du Bois Duddell apparently realized, indicated the production of a mishmash of audio-frequency waves. In 1899 Duddell discovered that placing a condenser in the air gap circuit caused the arc to hum at a more or less constant pitch. In effect, the condenser completed a resonant circuit because arcs already contained inductive choke coils to stabilize the heavy current flow. Eventually, Duddell found a way to control the pitch precisely enough to warble "God Save the Queen" on what he called his "singing," or "musical" arc.[15] One could plausibly argue that because each musical tone corresponded to a different wavelength, Duddell invented FSK and therefore a kind of frequency modulation. But he never concerned himself with telegraphy and, moreover, his instrument oscillated below 30,000 cps, well under the minimum threshold required for electromagnetic communications.

In September 1902 Valdemar Poulsen and P. O. Pedersen improvised three modifications that dramatically elevated the device's oscillation frequency: substituting water-cooled copper "beaks" for the electrodes; burning the arc in an atmosphere of compressed hydrogen (or a hydrogen-compound gas); and placing the arc in a strong magnetic field. To be sure, not even Poulsen understood why these changes caused his arc to radiate at radio frequencies, and a residual shushing sound betrayed the arc's imperfections as a sinusoidal generator.[16] But well after the advent of vacuum tube oscillators in 1913, the arc reigned as the best high-wattage emitter of continuous radio-frequency waves.

Poulsen saw frequency modulation not as a solution to a problem but only as a loathsome expedient—a necessary evil to tolerate until he could work out a means to amplitude-modulate his invention. The arc's heavy current was the chief obstacle to this goal. Starting up and keeping it going required a vigilant human operator to maintain as constant an amperage as possible in the antenna circuit. Dips and surges from amplitude-modulating the device risked causing the arc to shift its waves to another frequency or multiple simultaneous frequencies, or even to shut down. Attempts to change the antenna current also often resulted in dangerous and destructive high-amperage "secondary arcs" across the telegraph key's open terminals. Consequently, something as simple as sending Morse code by abruptly starting and stopping the current proved exceedingly difficult with even small arcs and impossible with large ones.

To get around this problem, Poulsen essentially replicated Ehret's FSK method. Rather than modulate the transmitter wave's amplitude, Poulsen alternated its length slightly by cutting a small value of inductance or capacitance in and out of an *LC* circuit. Unlike Ehret's invention, the Poulsen arc worked splendidly, but Poulsen emphatically objected to FSK on grounds of its profligacy with radio waves. Indeed, his condemnation of frequency modulation partly accounts for his lack of recognition for inventing the method. "The process of altering the length of the emitted wave," he insisted in 1906, "must be abandoned fundamentally, since this implies that each sending station would be characterized by two waves, and thus the number of stations which can work on the same service would be reduced to one half."[17]

His aversion to FSK notwithstanding, Poulsen used the technique at several Danish arc stations—again, as a temporary measure until someone worked out how to modulate the current's amplitude.[18] Naturally, Poulsen tried his own hand at this challenge. On one occasion, he claimed to have "a good method [where] the telegraph key throws the antenna and its balancing capacity in and out of connection with the other parts of the system, in which the oscillations are allowed to pass uninterruptedly."[19] He intended with this complicated scheme to isolate the high-amperage parts of the arc from the transmitter antenna, but one can scarcely imagine how the circuit could have accomplished this without interrupting the continuous wave and pitching the oscillator into an unusable state. Not surprisingly, no evidence exists for the widespread use of Poulsen's "good method" of amplitude modulation.

Poulsen's distress from using FSK was rendered moot in 1909, when he sold the arc's American patent rights to a recent Stanford University graduate from Australia named Cyril F. Elwell, who founded the Federal Telegraph Company in San Francisco a year later. Elwell and his engineers harbored no qualms about using FSK, which they justified on pragmatic grounds. The arc's "persistency," as Elwell flatly explained, makes it "irresponsive to rapid variations of current."[20] By and large, all of Federal's transmitters used circuits that resembled Poulsen's and Ehret's: a telegraph key switched an inductance in and out of the resonant circuit, which caused the circuit to alternate between two resonant wavelengths. Federal Telegraph receivers were perhaps even simpler than Ehret's slope detector, for they detected only the longer of two transmitted wavelengths. This method amounted to binary amplitude modulation of the longer "mark" wave, because a receiver wastefully ignored the shorter wave—the "space." Lee de Forest, who worked for Federal in 1913, spun this method as a security feature, because an eavesdropper who tuned to the redundant shorter wave heard instead of the normal pattern a confusing signal of transposed marks and spaces. Noting that amateur radio operators had complained about the difficulty of copying Morse code from such a "reversed signal," de Forest quipped that "we feel responsible for [their] state of thorough disgust."[21]

The experience of Ehret, Poulsen, and the Federal Telegraph Company engineers raises two questions about technological options. First, does their use of FSK—and the fact that Ehret and Poulsen invented FSK independently of each other—prove that the method was inevitable? Or did other choices exist? The inability of Poulsen and Elwell to amplitude-modulate arcs indicate they did not. So does the failure of Reginald Fessenden, who in a long-term quest to perfect the arc attempted to forge an entirely different path. In 1893 Fessenden's friend, and one of the founders of General Electric, Elihu Thomson, patented an arc that he claimed oscillated at "ten thousand, twenty thousand, thirty thousand, fifty thousand per second, or more"—barely above the minimum threshold of radio frequencies.[22] Nine years later, Fessenden reported that he "had by his experiments verified" Thomson's claim, and indeed in 1907 Fessenden, who apparently knew nothing of Poulsen's recent work, hailed "the genius of Professor Elihu Thomson for practically every device of any importance in this art."[23] But Fessenden never tried to frequency-modulate the arc and instead strove to make amplitude-modulation arcs work. Fessenden's employer, the National Electric Signaling Company, paid for his loyalty to AM by selling no more than a few low-power arcs—all amplitude-modulated radio telegraphs—and thus the firm failed even to approach Federal's success in that field. Partly for this reason, no historian to date has ever mentioned Thomson's and Fessenden's arcs in print.

By the end of World War I, radio practitioners recognized FSK as the de facto standard for all but small systems. Elmer E. Bucher, an RCA engineer who published a widely read radio engineering textbook in 1921, rejected binary amplitude modulation out of hand as impractical for arcs, and he saw no alternative to FSK: "It is obvious," he declared, "that a telegraph key cannot be placed in series with the arc gap for signalling [with amplitude modulation] and, in consequence, the formation of the Morse characters is usually effected by changing the inductance of the antenna circuit."[24] Moreover, FSK survives today. If Poulsen were alive today, he might be astonished to discover that the method of modulation he dismissed as a wasteful workaround remains in common usage, having outlived the arc itself by nearly a century.

One could also ask why no one before 1920 took the next logical step of placing Ehret's reactive microphone modulator in an arc to make an FM radio transmitter. Possibly the arc's characteristic hiss explains this inaction, but otherwise the technology was feasible, and Reginald Fessenden even came close to making it. In 1901 he mechanically coupled a microphone to reactive components in a resonant arc circuit, which produced "a change in the frequency or the natural period of vibration"—exactly the same technique for frequency modulation that Ehret and Poulsen independently invented a year later.[25] But Fessenden used his circuit to drive indirectly an experimental amplitude-modulation transmitter, not a frequency-modulation one. Some might interpret this as failure of imagination on Fessenden's part, but in 1901 he had no reason yet to give up on amplitude-modulation radio, which he was in the early stages of developing. After all, wireless telegraphy itself was only five years old.[26]

In July 1920 Alexander Nyman, an employee of Westinghouse, applied for the only patent ever issued for an arc-based FM radiotelephone system (fig. 11). In terms of modulation, Nyman invented nothing new. He fitted a Poulsen-like arc with an Ehret-style microphone modulator. The receiver was updated, though; it was a vacuum-tube circuit that his patent tersely described as a "simple receiving station" and illustrated with a drawing of something resembling an Ehret detector of 1902, showing that the slope detector was part of normal practice by 1920.[27] Indeed, it is difficult to see what Nyman hoped to achieve when one notes that the arc was already fast approaching obsolescence because of newly available, cheaper, and more reliable vacuum-tube oscillators. Nor did he disclose any advantage of FM radiotelephony over AM. Although Nyman's electronic invention would likely have worked far better than Ehret's radiotelephone, Westinghouse engineers likely perceived no urgency to develop an alternative to AM radio. Thus, Nyman's invention had almost as negligible an effect as Ehret's, and its chief historical significance is to reveal—again—that a heavy layer of conservatism can often underlie even unorthodox ideas.

FM radiotelephony descended into a state of moribundity from 1902 until 1920, but thanks to the Poulsen arc, FM radiotelegraphy thrived as normal practice during the same period. As the widest-used continuous-wave radiator until after World War I, the FSK-modulated arc demonstrated the practicality of altering the wavelengths of radiators with virtually unlimited wattage. In the final analysis, Poulsen's and Elwell's radio*telegraphs* cleared a far wider and more direct path to FM radiotelephony than did Ehret's and Nyman's radio*telephones*. Howard Armstrong once said as much to Cyril Elwell himself. In October 1940, when Armstrong was giving a speech about modern wideband FM before the American Institute of Electrical Engineers, he recognized Elwell sitting in the audience. Armstrong pointed him out and introduced him as "one of the first users of frequency modulation in the days of the mark-and-space keyed arc transmitters."[28]

The Crystal Detector and the Rise of Amateur Radio Clubs

Nothing in the social history of radio accelerated the development of FM radio more than the invention of the crystal detector in 1906. Before that year, the expense of apparatus like the coherer made even the reception of wireless messages chiefly the province of well-funded corporations, entrepreneurs, and

Jan. 25, 1927. 1,615,645

A. NYMAN

COMBINED WIRELESS SENDING AND RECEIVING SYSTEM

Original Filed July 15, 1920 2 Sheets-Sheet 1

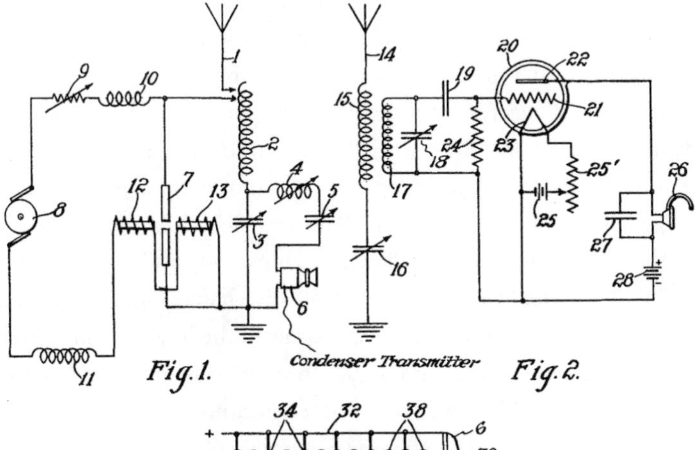

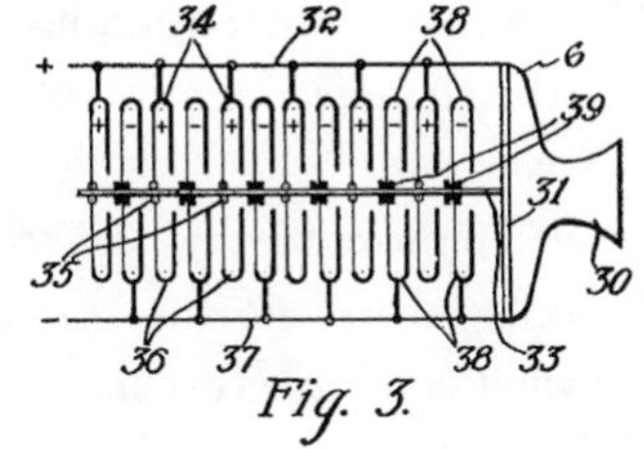

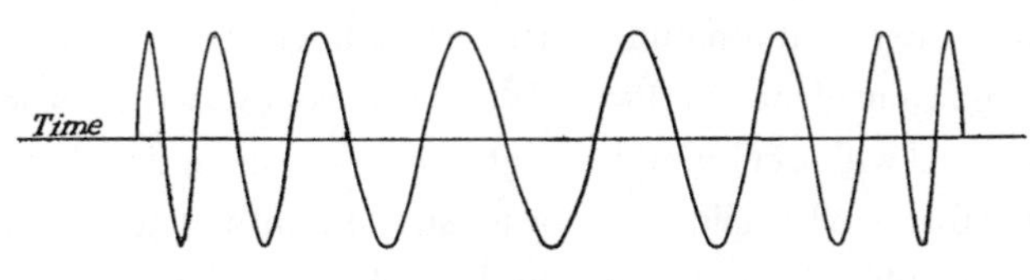

Fig. 4.

WITNESSES:
H. J. Shelhamer
A. Press

INVENTOR
Alexander Nyman
BY
Wesley G. Carr
ATTORNEY

Fig. 11. Nyman FM Radiotelephone Patent, 1920. Nyman FM is based on the oscillating arc. Alexander Nyman, "Combined Wireless Sending and Receiving System," U.S. Patent No. 1,615,645, application date: 15 July 1920, issue date: 25 January 1927, assigned to Westinghouse.

governments. Afterward, the price of radio receivers plummeted, swelling the ranks of wireless practitioners with hundreds of thousands of hobbyists. Many were boys who grew up to be radio engineers; virtually every FM inventor of any importance started out as a young amateur radio operator.[29]

The crystal detector transformed the technological context of wireless communications in a matter of months. In late November and early December 1906, two Americans, Greenleaf Whittier Pickard and Henry C. Dunwoody, were independently issued patents for essentially the same device: a circuit that took advantage of the peculiar electrical properties of ordinary crystalline minerals such as silicon, galena, and carborundum.[30] Although they had not worked out the physics behind their inventions, Pickard and Dunwoody understood that crystals rectified—that is, filtered out all but the positive halves—of a high-frequency radio wave. From the rectified wave was extracted the lower-frequency components; namely the sound waves and telegraph signals superimposed on the high-frequency radio waves. Moreover, the crystal detector exhibited a sensitivity and a stability far superior to the much more expensive coherer, required no external source of power, and never wore out.

The replacement of the coherer with the crystal detector created social effects on a scale that dwarfed the technical ones. Because crystals cost about a dollar, the expense and complexity of wireless receivers declined precipitously.[31] A community of amateur radio operators—hams, they called themselves—arose and established a tradition of camaraderie and technological enthusiasm unmatched until the advent of the personal computer. That wireless fascinated so many boys, and not girls, was no coincidence, for Pickard's and Dunwoody's technological transformation meshed with a cultural shift that was already in motion. Susan Douglas has written that wireless allowed a boy to "straddle old and new definitions of masculinity." The older "primitive" ideal valued physical strength, a "commanding personality," and direct contact with nature. In contrast, Douglas says, a new masculinity emerged from the recent urbanization, corporatization, and mechanization of American society. Opportunities for outdoor experiences—especially for city youths—diminished, and traditional manly values were devalued. More and more, intelligence, education, and specialized technical knowledge were seen as opening paths to successful lives and careers. Although building and operating a radio station seldom called for outdoor or strenuous activity (except, perhaps, during the often-precarious job of erecting an antenna mast), by manipulating electromagnetic waves, hams demonstrated mastery over one of the most mysterious phenomena of nature.[32]

From the beginning, no institution shaped and spread the technology and cultural values of amateur radio more than did local, regional, and national radio clubs. Although neither the number of clubs nor their total membership is known precisely, contemporary sources indicate considerable early growth and numbers. In early 1908 the magazine *Electrician and Mechanic* founded "The Wireless Club," and by September that organization boasted chapters in 114 cities and towns in the United States and Canada.[33] A year later the energetic editor, publisher, science fiction author, and mail-order entrepreneur Hugo Gernsback created the Wireless Association of America, which by February 1913 claimed 230 affiliates.[34] In January 1910 the *Outlook* magazine estimated that more than 4,000 amateur radio operators lived in the United States.[35] Three years later, the Radio League of America counted some 350 local clubs and more than 300,000 radio amateurs in the United States altogether.[36]

Radio technology fostered especially well what has been seen as a culture of "brotherhood" or "fraternity" among radio amateurs, principally by connecting them with distant fellow hobbyists. Along with magazines, mail-order stores played a major role in expanding the worldview of many a young amateur. Because few firms before the 1920s marketed a completely assembled receiver, virtually all listeners built their own sets with mail-order parts. Hugo Gernsback, in New York City, operated the most important store in America, the Electro Importing Company, for this purpose. Boys all over America relied on Gernsback for how-to articles and radio parts, advertised in his *Modern Electrics* magazine. Harold Beverage, an engineer who participated in RCA's earliest experiments with frequency modulation during the 1920s at the company's Riverhead, New York, laboratory, credited Gernsback with his initial exposure to the field of radio. In a 1968 interview, Beverage recalled that, as a youth in rural Maine, "I got interested in a magazine called *Modern Electrics*. It was put out by one Hugo Gernsback. . . . That was quite interesting to me, fascinating, so I sent away and got a catalog from [Gernsback's] Electro Importing Company. . . . I bought [a condenser] and made my own coils. I swiped a piece of galena [crystal] from the high-school laboratory."[37]

The excitement of ham radio ruined Beverage for farming. "I used to copy a lot of news from a station on Cape Cod . . . which was sending out news to the ships at 10 o'clock at night," he recalled. "Back on the farm I thought it a lot more fun to be messing around with wireless than it would be pitching hay." Harold Peterson, a Nebraskan who partnered with Beverage at RCA during the 1920s, also credited Gernsback with his introduction to radio. *The Electro Import Catalogue*, he said in an interview, "had a nice little description of radio, how it works and what it could do. I remember reading that over and over again, and got started that way." Peterson, whose family also lived on a farm, "DX'd" stations as far away as Washington, D.C.[38] Countless radio engineers during the twentieth century could tell similar stories.

Among amateur organizations, the Manhattan-based Radio Club of America was the most closely associated with the origins of modern FM radio. The group began to take shape in 1907, when "three small boys"—George Eltz Jr., Frank King, and W. E. D. "Weddy" Stokes Jr.—met to fly model airplanes. Initially they christened their gang "The Junior Aero Club of America" and elected eleven-year-old Weddy president. Discussions at gatherings turned more and more to wireless, however, so the boys briefly called themselves "The Wireless Club of America" before permanently settling on "The Radio Club of America."[39] Today, a full century later, the Radio Club of America still exists; no other organization in the world has dedicated itself solely to radio for a longer continuous period of time.

The earliest members of the Radio Club of America by and large belonged to middle-class or relatively prosperous families, which enabled them to purchase more expensive apparatus. Armstrong, who joined about 1912, was the son of the American representative of the Oxford University Press. Weddy Stokes descended from a wealthy family of shippers, and his father was a successful racehorse breeder and entrepreneur who built and owned the deluxe Ansonia Hotel. The elder Stokes encouraged his son's interest in radio and allowed Weddy to host club meetings and to install in the Ansonia a wireless station with a 10,000-watt transmitter. Because the rig required an immense amount of power—more than all but a handful of military and commercial stations possessed at the time—keying the transmitter overloaded the hotel's in-house electrical generator, provoking guests to complain about flickering lights in their rooms.[40]

During its earliest years, the Radio Club forged three hallmark amateur radio traditions that would shape FM radio technology and the way it would be promoted: public demonstrations that showed off the social usefulness of radio, political activism, and a fellowship among amateurs that often transcended commercial interests. One of the club's first demonstrations occurred in 1913, when charter members Frank King and George Eltz constructed a small, doubtlessly low-fidelity arc radiotelephone transmitter in King's home on West 107th Street. Ten years later a reporter cited this project as "one of the first radio telephone broadcasting stations in the United States." It might also have been among the most dangerous to operate. Because the arc burned in a pressurized chamber of inflammable vapor, the boys had to synthesize a supply of gas by heating alcohol over a flame. Sometimes the "all home made, and naturally crude" apparatus spontaneously ignited. "Several amusing incidents occurred when the mixture in the arc chamber became explosive," the same reporter wrote, "and the operators were forced to beat a hasty retreat." Frank and George escaped these and other perils to broadcast phonograph records for several navy "battleships swinging at anchor a short distance away in the Hudson River."[41] Two years later, the club's Ansonia Hotel station relayed more than a thousand telegraph messages for the navy during another port of call. The club pulled off its grandest demonstration in 1921, when its members exchanged the first transatlantic messages using short waves. Howard Armstrong was one of five stateside radio operators in this experiment, and his friend Paul Godley, who would help publicize the Armstrong system of high-fidelity FM during the 1930s, operated the club's station in Scotland.[42]

The Radio Club also excelled in political activism. Indeed, the amateur movement as a whole began largely as a populist uprising. Gernsback founded the Wireless Association of America in 1909 "in order to guard against unfair legislation as far as the wireless amateur was concerned."[43] In 1910 Gernsback marshaled the collective force of amateurs to resist the first proposed legislation that was harmful to the interests of hams. "The association had no sooner become a national body," he reminded his readers in 1913,

> than the first wireless bill made its appearance. It was the famous Roberts Bill, put up by the since defunct wireless "trust." The writer [Gernsback] single handedly, fought this bill, tooth and nail. He had representatives in Washington, and was the direct cause of having some 8,000 wireless amateurs send protesting letters and telegrams to their congressmen in Washington. The writer's Editorial which inspired the thousands of amateurs, appeared in the January, 1910, issue of *Modern Electrics*. It was the only Editorial during this time that fought the Roberts Bill. No other electrical periodical seemed to care a whoop whether the amateur should be muzzled or not. If the Roberts Bill had become a law there would be no wireless amateurs to-day.[44]

No individual local club defended the interests of hams more than the Radio Club of America did. In April 1910 George Eltz, Frank King, Weddy Stokes, and Ernest Amy traveled to Washington to lobby against another piece of proposed legislation known as the Depew Bill, which if enacted, the boys feared, would prohibit amateur transmissions. Fourteen-year-old Weddy testified before the Senate Commerce Sub-Committee, delivering, by one account, an appealing call to alarm. "Clad in knickerbockers," the *New York Herald* reported, "he captured the hearts if not the judgment of the Senators." Weddy denounced the Depew Bill as a "stock-jobbing scheme," and warned that "soon a great trust will be organized to corner the very air we breathe."[45] Two years later the Radio Club dispatched a second delegation to Congress to lobby against a similar bill.[46] In 1922 the club's president, Howard Armstrong, represented American amateur radio operators at the first of four annual National Radio Conferences, hosted by Secretary of Commerce Herbert Hoover. These meetings marked Armstrong's debut as a "public engineer" whom government officials consulted about radio-related issues. During the remainder of his life he testified several times as an expert witness before congressional committees and the FCC, most often during the 1930s and 1940s as an advocate of wideband FM radio.

One of the greatest strengths of the Radio Club of America was its tradition of welcoming all radio practitioners, both amateur and professional. Club policy encouraged members to share technical and other information, a practice that blurred proprietary boundaries and facilitated transfers of knowledge. This was not a unique role for the club; in 1913 the all-professional Institute of Radio Engineers was established for similar purposes, and the two groups shared many members. But no other organization merged so seamlessly the camaraderie of ham radio with a professional-like seriousness of purpose that sometimes led to first-rate research. Admittedly, papers that appeared in the *Proceedings of the Radio Club of America* as a whole lacked the theoretical rigor one could expect to find in the *Proceedings of the IRE*, but the Radio Club possessed an unrivaled atmosphere of "fellowship." As *Radio Broadcast* magazine explained in 1923,

> A club is a place for good fellowship, true; and that describes the Radio Club of America, which has already stimulated good fellowship in radio and more specifically among its members. In that sense, the word stands.
>
> But in the case of this group of young men, there has been something more than a club atmosphere. With the serious intentions of its members, the thoroughness of the papers and discussions marking its meetings, and the scientific value of its experiments and tests, the word "club" is almost a misnomer. This organization might well call itself a scientific society, although it does retain that spirit of fellowship which goes with the usual meaning of club.[47]

By the mid-1920s, according to club historian George Burghard in 1934, most of "the original small boys had grown to be full fledged men of affairs" in the radio industry. "Naturally," Burghard explained, "the character of the membership of the club as-well as that of the papers, underwent a similar change. The club had now all the earmarks of a genuine scientific body. The spirit of the organization, however, never changed. These men, now engineers, executives or scientists were still amateurs at heart."[48] Such a description fit Armstrong perfectly. In his entire life, he held only two salaried jobs: his military service during World War I, and a dollar-a-year research professorship, which Columbia University gave him during the late 1920s.

Traditions of technological expertise, openness, and friendship among practitioners largely accounted not only for how widely word about experiments with FM radio spread through the radio industry during the 1920s but also for why both amateurs and professionals participated in developing, testing, and promoting Armstrong's wideband FM during the 1930s. Five friends of Howard Armstrong—Tom Styles, Jack Shaughnessy, Paul Godley, George Burghard, and especially Carman Runyon Jr.—all long-standing members of the Radio Club—helped test and sell the Armstrong system to the public, sometimes for no remuneration. Styles worked full time as Armstrong's financial manager and secretary, Shaughnessy as his assistant; in 1936 Godley published the first lengthy article explaining Armstrong FM to the broadcasting industry; and in 1934 Armstrong installed a prototype receiver in Burghard's Westhampton Beach, Long Island, home during RCA's initial field tests of the Armstrong system.[49] Runyon, a founding member who managed a coal-delivery company in Yonkers full time, witnessed Armstrong's secret work with FM as early as 1932 and for several years afterward routinely took part in public demonstrations of broadcast FM radio.[50] Other members involved themselves with experimental broadcast FM, sometimes in connection with Armstrong's RCA trials, sometimes with the FM work of other companies, sometimes as pioneer FM broadcasters. Harold Beverage, who pulled strings to obtain permission for Armstrong to test his FM system in RCA's Empire State Building transmitter in 1934, joined both the Radio Club and the Institute of Radio Engineers soon after World War I. Harry Sadenwater, an RCA Manufacturing Company engineer, permitted Armstrong to install prototype receivers in the basement bar of his home in Haddonfield, New Jersey, in 1934. John V. L. Hogan, a veteran wireless pioneer who had worked for both Fessenden and de Forest, founded WQXR, the first broadcast FM radio station in Manhattan, in 1939. Albert Goldsmith, who was also a former president of the IRE, evaluated Armstrong's FM system for RCA in 1934. Beverage's colleague at RCA Communications Company, Murray Crosby, patented several FM-related inventions before Armstrong did, published important technical papers about frequency modulation, and attended Radio Club meetings as a guest from time to time. One man, Howard Armstrong, patented wideband FM in 1933, but a community of both professional and amateur practitioners also contributed to the development of the technology.

Such was the technological and social context of radio from 1902 until the early 1920s, a context out of which emerged both AM and FM broadcast radio. Certainly, frequency modulation was nothing new by the time Armstrong's patents were issued in 1933; FM thrived in radiotelegraphy, and minimal imagination was required to begin adapting the method for sound. Even more important to the progress of FM was the culture surrounding radio technology that formed years before 1920, for the final shape of radio owed much to a radio engineering profession that an amateur tradition deeply influenced.

Congestion and Frequency-Modulation Research, 1913–1933

I'm forever losing signals,
Pretty signals in the air;
They're pitched so high,
Nearly reach the sky,
Then like my dreams they fade and die,
Signals always fading,
I've tuned everywhere.
I'm forever losing signals,
Pretty signals in the air.
(sung to "I'm Forever Blowing Bubbles")

Lose M. Ezzy, 1920

We have worked intermittently for a long time on what we call "modulated frequency" which has many advantages over the present method.

Harry P. Davis, President of Westinghouse, 1931

Congestion and the Creation of the Spectrum Paradigm

For nearly twenty years after Cornelius Ehret and Valdemar Poulsen filed their patent applications in 1902, frequency-modulation radiotelephony languished in the backwaters of radio engineering as something to ponder from time to time, but ultimately dismissed as impractical or unneeded, or both. After World War I, though, more and more practitioners took up FM, thanks chiefly to the broadcasting boom that began in 1920 and the problem of congestion that followed.

Historians of radio often characterize broadcasting in the early and mid-1920s as a period of ever-worsening "congestion" and "chaos." Because the federal government lacked the authority to regulate the new industry effectively, broadcast signals began overcrowding the airwaves.[1] The number of licensed stations surged from a handful in late 1920, to 28 in January 1922, to 570 in December of the same year, prompting many observers to fear that the very success of broadcasting would soon suffocate the new medium.[2]

The rise in the number of stations only hinted at the peril radio broadcasting faced. The increased popularity of radio aggravated many old ills. Sporadic fading worsened as stations attempted to reach more distant listeners. Lightning storms frequently ruined local AM reception, and solar flares halted radio communications worldwide. Station transmitters ratcheted up in wattage, which increasingly disrupted reception, because the radiation patterns of formerly distant stations tended more often to overlap. Moreover, until the mid-1920s no station could precisely stabilize the wavelength of its carrier because changes in temperature and humidity altered the resonant frequency of an *LC*-controlled oscillator circuit.[3] In one of the most notorious cases of carrier drift, listeners in the Dearborn, Michigan, area might have found the Ford Motor Company's station anywhere between 800 and 980 kHz, a range spanning nineteen modern-day AM channels.[4] Finally, the electrification of America continuously layered on new strata of noise, originating chiefly from the sparking motors of household appliances. As one listener complained, radio was plagued by the "all too-familiar buzzes, crackles and frying occasioned by your own or your neighbor's electric razor, oil burner, kitchen mixer or vacuum cleaner."[5] In fact, anything that created sparks, including hospital X-ray machines, automobile spark plugs, and trolley cars, could spoil reception.

At first, averting congestion seemed merely a matter of allocating more wavelengths. In 1920 the Commerce Department's Bureau of Navigation, the agency to which Congress gave modest powers in 1912 to oversee American radio, declared that all broadcasting stations must use the same 30-meter wavelength. This arrangement sufficed so long as broadcasters remained small in number, low in power, and separated by adequate distances. But no law allowed the Navigation Bureau to limit the number and wattage of transmitters; as more and more stations went on the air, the agency was forced within a year to assign second, third, fourth, and fifth wavelengths. Each new allocation, though, proved increasingly less efficacious than the preceding one, and by the mid-1920s American broadcasting had indeed descended into chaos. Not until 1927 did Congress address the problem, when it created the Federal Radio Commission (FRC), a panel empowered not only to thin out the tangle of broadcast stations but also to determine the wattage, carrier frequencies, and hours of operation of those remaining on the air.

To do its work the FRC adopted a new paradigm based on the concept of the radio spectrum. The most apparent sign of this change appeared in the vocabulary of practitioners. Before and during the early years of the broadcasting boom, one spoke of a tuning to the *wavelength* of a transmitter and never to its *frequency*. The Radio Act of 1912 used the term wave length to the complete exclusion of frequency, requiring, for example, that ship stations "use two sending wave lengths, one of three hundred meters and one of six hundred meters."[6] By contrast, after 1927, one almost always spoke of the position of a transmitter's carrier wave with respect to the electromagnetic spectrum—that is, the carrier's frequency.

During the 1920s, the FRC rationalized the job of diluting congestion by using the spectrum as a one-dimensional map that graphically represented bands of radio frequencies in the same way that two-dimensional geophysical maps symbolize land. The standard AM broadcast band, the best-understood and stablest "territory" of the spectrum, extended from approximately 550,000 to 1,500,000 cps. Above that lay the ill-defined, continually upwardly shifting boundary that marked the frontier, and beyond that the "short waves" and "ultra-highs," newly discovered expanses of the spectrum where the FRC (and later the FCC) would allocate for television and other radio services. Evocative of the land-grant offices handing out acreage to American settlers during the nineteenth century, the language of the new radio landscape acquired its "pioneers," and "land rushes"—grabs for recently opened parts of the spectrum.

The spectrum paradigm not only explained interference and congestion but also helped the FRC partly solve those problems. Two stations, for example, each one restricted to a 10-kilocycle-wide channel, would not interfere with each other if the FRC assigned them carrier frequencies separated by at least 10,000 cycles per second. But a receiver designed for 10-kilocycle channels would detect stations with overlapping channels. For remedies, the FRC could choose among several, including shifting the channel assignment of one of the transmitters, reducing the power of one or both stations, or withdrawing one of the broadcasters' licenses.

Although by 1930 everyone who worked with radio took the spectrum for granted, almost no one did seven years earlier. Practitioners made do instead with analogies that implicitly compared radio to older technologies, such as the telegraph or telephone. In fact, the very invention of radio had hinged on using the right analogy. During the 1890s, a twenty-one-year-old Guglielmo Marconi invented the wireless telegraph, beating out several scientists who sought at the time a practical use for electromagnetic waves. The professionals, as historian Sungook Hong has explained, fixated on an optical analogy, which caused them to overemphasize the similarity of radiofrequency waves to light waves. This thinking prevented them from imagining that EM waves could send messages over great distances. By contrast, Marconi borrowed his analogy from overland telegraphy. His coinage of the term wireless telegraph indicates that he saw his invention as a wireless version of an older long-distance communications technology.[7]

The wireless telegraph (along with the wireless telephone) was the most common analogy used to explain radio during its first quarter century, but another deserves mention. Lee de Forest, inventor of the audion, the earliest electronic amplifying device, boasted in a 1916 magazine article that he was the first to use his invention "as a printing press." De Forest observed that the electromagnetic newspaper "has the further great advantage that it can be delivered instantly and without the nerve-racking cry of extra, in the quiet of your home, without opening the door, or even ringing your bell."[8] Others had been using a printing press analogy for electrical communications in general for decades. In 1887 Edward Bellamy described a system of wired telephone broadcasting in his utopian novel, *Looking Backward,* and reports of a "telephonic newspaper" introduced in Hungary in 1893 appeared in the United States the same year.[9] Furthermore, a Newark, New Jersey, company was transmitting a "telephone newspaper" to subscribers four years before de Forest's article about radio broadcasting.[10] After 1920, though, the radio-newspaper analogy began to fall apart because radio congestion resembled nothing in the experience of brick-and-mortar publishing.

Congestion alone forced the creation of the spectrum paradigm. When the broadcasting boom began, the federal government was still operating under the Radio Act of 1912, enacted during an age of spark gaps, coherers, damped waves, and point-to-point radiotelegraphy. Within a few years, the government's regulatory authority began to lag behind major changes in the technology, such as the mass production of electronic vacuum tubes. Those who worked with radio realized that something had to be done. In early 1922, as part of a publicity campaign to prod Congress to give the government greater regulatory powers, Secretary of Commerce Herbert Hoover invited thirteen leaders from various organizations—amateur radio clubs, the Institute of Radio Engineers, government agencies that used radio, and the radio manufacturing industry—to attend a conference about the problem, in what was to be the first of four annual meetings. Building on its tradition of public service, the Radio Club of America dispatched Howard Armstrong as the presumptive representative of ham radio operators.[11]

Delegates to these conferences sketched out several ideas that the FRC and, later, the Federal Communications Commission eventually adopted. The first group of attendees, although by no means united on every issue—amateurs, for example, feared an AT&T broadcasting monopoly—by and large agreed on the need to limit the number and power of transmitters. In April 1922 K. B. Warner, editor of *QST*, the leading American amateur radio magazine, warned that "in recent months the radio game has progressed to a point where it simply cannot wait any longer for new regulations." He reported an alarming—and accurate—rumor that "some five hundred applications for broadcasting [are] pending in the Department of Commerce." "Everybody [cannot] be wholly satisfied simply because there aren't enough wave lengths," he observed. Warner urged Congress to grant "the Department of Commerce wide discretionary powers, with the authority to issue, amend or revoke regulations and licenses." Warner also hoped that the Commerce Department would regulate "so as to be of the greatest good to the greatest number of our people."[12]

At this point the spectrum paradigm began to coalesce. Before 1922, few outside physics laboratories had even mentioned the spectrum, but the men who attended the first conference early that year used language that reflected how much they had integrated the spectrum into their thinking. Two months before hosting the first radio conference, in April 1922, Hoover struggled to find words to explain congestion, observing that the proposed regulation of radio was "rapidly becoming as vital a topic as forest preservation and protection of water power rights." The "air is full of chatter," he declared, and with only four or five wavelengths, "ordinary wireless telephonic communication . . . has clogged up this medium of communication to such an extent, that . . . some form of 'ether cops' will have to be established to regulate traffic." Hoover's reference to "wavelengths" evidenced the persistence of the old wireless telegraph analogy, and he borrowed "chatter" from amateur radio slang. But his comparison of radio with natural resources such as forests and water power reveals that he was beginning to conceive of radio in terms of a virtual landscape, similar to the one that the spectrum paradigm would map out.[13]

A year later, further shifts in language signaled a complete transformation. In March 1923 Commissioner of Navigation David B. Carson opened the second conference by asking the fifty attendees to "confine yourselves strictly to broadcasting and the allocation of wavelengths." A few days later, the assembly, after considering solutions to congestion, advised that the government abandon the practice of assigning individual wavelengths in favor of a new broadcast band from "222 meters to 545 meters, with the 'government reserve' above 600 meters opened up to take care of some of the displaced services."[14] This recommendation retained the older idiom, but the group also first invoked the spectrum paradigm explicitly by suggesting "that radio stations be assigned specific wave *frequencies* (wave lengths) within the wave *band* corresponding to the service rendered."[15] Indeed, topping the list of the conference's "more interesting and important resolutions," according to *QST*, were proposals to assign "a wave band of 10,000 cycles to each Class A broadcasting station" and to space geographically proximate stations no closer to each other than 20,000 cycles, with 50,000 cycles between distant stations. The Commerce Department speedily adopted most of these recommendations and was soon assigning stations to standardized 10-kilocycle channels.[16] This action marked when an authoritative body first embraced, albeit informally and probably unconsciously, the spectrum paradigm. The Bureau of Navigation made the conversion permanent when the agency's *Radio Service Bulletin* announced in May 1923 that it would begin publishing station frequencies, with equivalent wavelengths in parentheses. On the fifteenth of the same month, however, the *Bulletin* officially stopped listing wavelengths altogether.[17] Since the late 1920s, the FRC and later the FCC have employed the spectrum paradigm as an intellectual framework for analyzing and alleviating the problems of interference and congestion.

Eighty years ago, however, regulation offered no panacea for the ills of broadcasting. American critics, especially, protested that a system of licensing stations conflicted with traditions of free speech. Soon after Congress created the FRC in 1927, one new commissioner described his job of determining "who shall and who shall not broadcast" as "a rather appalling responsibility. The law tells us that we shall have no right of censorship over radio programs, but the physical facts of radio transmission compel what is, in effect, a censorship of the most extraordinary kind." He pointed out that broadcasting resembles in some ways the newspaper business, "but with this fundamental difference[:] there is no arbitrary limit to the number of different newspapers which may be published, whereas there is a definite limit, and a very low one, to the number of broadcasting stations which can operate simultaneously within the entire length and breadth of our country."[18] As some historians of radio regulation have observed, the first few years of the new regulatory system bogged down in controversy, when business interests with allies on the FRC and FCC muscled out most radio stations owned by nonprofit organizations.[19] Not surprisingly, because regulation alone could do only so much to solve congestion, a number of technological proposals also emerged during the 1920s and 1930s.

The Revival of FM Radiotelephony Research

Technological fixes for congestion included a motley collection of "static eliminators" and "static reducers," nonfunctional contraptions that usually originated in the workshops of freelance inventors. A few ideas worked splendidly, though, nearly all of which were inconceivable without the spectrum paradigm. Armstrong's superheterodyne circuit markedly improved receiver selectivity—the ability to tune in a single radio channel and mask adjacent ones—which in turn allowed more stations on the air.[20] George Pierce patented the crystal oscillator, which replaced traditional inductors and condensers in a resonant circuit, with a small piece of vibrating quartz, to attain far more stable carrier frequencies. This device permitted the FRC to pack stations more tightly into the broadcast spectrum.[21] AT&T engineer John R. Carson invented single-sideband modulation (SSB), which took advantage of the fact that the two sidebands in an AM channel carry redundant information. By suppressing one sideband, Carson halved the width of an amplitude-modulation channel, thereby theoretically doubling the maximum number of stations permitted on the air and halving static noise levels.[22]

Some practitioners also, for a few years, proposed a type of "narrowband" frequency modulation as a solution to congestion. The theory for the method derived from the fact that a normal amplitude-modulation signal requires a channel, a slice of radiofrequency spectrum wide enough to carry two mirror-image "sidebands." Although comprising a snarl of continuously changing audio-frequency waves that defy easy analysis, sidebands figure in a few straightforward principles for AM. First, each sideband requires a contiguous portion of spectrum, as wide as the audio-frequency bandwidth, to convey information. Thus, the channel carrying both sidebands spans *at least* twice as much spectrum as the audio bandwidth:

$$\text{AM channel width} \geq 2F_a,$$

where F_a represents the width of the audio bandwidth and of each sideband. A second principle states that widening a channel opens further the door to static noise. All this leads to a delicate balance between audio fidelity, which requires wider channels, and static noise, which tends to increase with channel width. Moreover, widening the standard channel width decreases the number of stations that can broadcast simultaneously across a fixed portion of the radiofrequency spectrum. Since the early 1920s, most engineers have accepted this matrix of trade-offs as inherent in normal AM radio practice.

A number of visionaries, though, questioned whether the same rules must apply to frequency modulation in the same way. Indeed, must the rules apply at all? Perhaps FM would sever the connections between fidelity and static and between static and the channel width. To understand how, imagine an FM transmitter that emits at any instant a single wave, not a band of mixed frequencies, as with AM. That wave's frequency "wobbles" from side to side across a small portion of the spectrum, the range of wobble proportional to the instantaneous amplitude of the modulating sound wave. Accordingly, an FM signal requires—one would hope—a channel wide enough to accommodate only the maximum extent of the wobble, called the frequency swing. Mathematically, it was simple:

$$\text{narrowband FM channel width} = 2f_m = 2hA_a,$$

where f_m is the maximum deviation from the center transmitter frequency (or half the frequency swing), A_a is the maximum audio amplitude, and h is the designer-chosen modulation index, a constant value. Because the swing varied with only the *amplitude* of the audio signal—in a constant proportion (h) selected by human designers, not by nature—then man, not nature, would determine the width of the channel. By 1923, with ever more stations jostling for channels on the spectrum, nothing would have been more propitious for broadcasting than a working system based on this theory. A designer could conceivably choose a modulation index h as small as the practical limits of electronics technology allowed, pack into the spectrum far more FM channels than the old AM method permitted, and obtain better audio fidelity to boot. Narrowing the channels would also shut out a great deal of static. At least, advocates of the theory hoped so.

The first published record of narrowband FM technology exists as a patent application filed in 1923.[23] But gossip had it that Frank Conrad, the founder of the pioneer broadcast station, KDKA, in Pittsburgh in 1920, experimented with narrowband at least as early as 1921. RCA engineer Clarence Hansell recalled eleven years later that Conrad "at one time advocated the use of frequency modulation in radio broadcasting." "Dr. Conrad," Hansell explained, "was credited with statements to the effect that frequency modulation did not produce side frequencies as in amplitude modulation and would therefore permit a great increase in the number of broadcasting stations."[24]

Conrad's claims collapsed when in February 1922 John Carson of AT&T published the first mathematical analysis of modulation in a historic article titled

"Notes on the Theory of Modulation." "A great deal of inventive thought has been devoted to the problem of narrowing the band of transmission frequencies," Carson declared in an early paragraph, adding that his article aimed "to analyze the more ingenious and plausible schemes which have been advanced to solve this problem." A page or two later he repeated this point, in a sentence remarkable for its twin coinage of the terms amplitude modulation *and* frequency modulation. "In order to eliminate the necessity of sidebands," he stated, "it has been proposed a number of times to employ an apparently radically different system of modulation which may be termed *frequency* modulation as distinguished from *amplitude* modulation, in the belief that the former system makes possible the transmission of signals by a narrower range of transmitted frequencies." "This belief is erroneous," Carson asserted, although he allowed that "the reasoning on which the supposed advantage [of narrowband FM] is based is very plausible."[25]

Carson's theoretical FM transmitter was not complex. He asked his readers to imagine a constant-frequency sinusoidal audio wave $sin(pt)$, which oscillates relatively slowly, say, at 100 cps. This modulates the frequency of a continuous-wave oscillator with a much higher carrier frequency ω_o, at perhaps 1 million cps:[26]

$$\omega = \omega_o\,[1 + h\,sin(pt)],$$

where ω is the instantaneous radio frequency emitted by an FM transmitter, ω_o is the center (unmodulated) radio frequency, h is the modulation index, and $sin(pt)$ is the hypothetical sinusoidal sound wave.

The preceding equation describes the simplest kind of frequency modulation: when a low-frequency audio wave $sin(pt)$ modulates a high-frequency carrier wave ω_o. This equation can be rearranged algebraically to clarify its meaning:

$$\omega = \omega_o + (h\omega_o)sin(pt),$$

which is equivalent to:

instantaneous frequency = center frequency + instantaneous frequency deviation

Ordinarily, this equation describes any kind of FM system, but not in Carson's article. Significantly, he restricted his analysis to systems in which the modulation index "h is small compared with unity." The purpose of this stipulation becomes clear when one observes that the maximum magnitude of $sin(pt)$ is ± 1.0, and thus the maximum magnitude of the frequency deviation is $\pm h\omega_o$ (i.e., the maximum value of $h\omega_o sin\,pt = h\omega_o \times [\pm 1.0] = \pm h\omega_o$). In other words, the small value that Carson assumed for h makes the maximum frequency deviation ($h\omega_0$), and therefore the channel width, small as well. Carson stressed that he chose a small value of h to conform to the premise of those who hoped narrowband FM would conserve spectrum.[27]

Following Carson's analysis in detail requires mathematical knowledge beyond the scope of this book, but understanding his conclusions in a general way does not. Contrary to the expectations of narrowband proponents, he argued, narrowband FM *does* produce sidebands; even worse, choosing a small h causes multiple sidebands to splay across a range of spectrum that far *exceeds* the channel width of a comparable AM signal. The term narrowband FM, therefore, refers to an impossibility. Furthermore, attempts to employ FM to narrow the channel width will actually *increase* audio distortion.

Paradoxically, although communications engineers have long regarded Carson's paper as a classic in their field, few radio engineers have suffered more unjustly at the hands of historians than John Carson. Lawrence Lessing, for instance, damns Carson's article for being "so injudicious as to draw the sweeping conclusion: 'This type of modulation inherently distorts without any compensating advantages whatever,'" asserting also that Armstrong liked to remind Carson of his "bloomer."[28] As Lessing's statement indicates, though, he and those who have followed his example have misconstrued Carson's analysis as an attack on *all* kinds of FM, something Carson clearly took pains not to do. In fact, Carson shared with Armstrong a flair for explaining clearly the gist of a highly technical argument. Even if a reader misses the fact that Carson chose a small modulation index h in order to evaluate only narrowband FM, she will find that Carson expressed no opinion about the feasibility of every type of FM, especially the not-yet-invented Armstrong kind.[29] Carson's critics have also explained his "mistake about FM" by implicitly caricaturing him as an out-of-touch egghead, a creature of theory, not practice.[30] This is a curious misrepresentation of Carson's career, for although he ranks among the greatest radio communications theorists of his time, few could match his practical accomplishments. In 1924, for example, the Institute of Radio Engineers awarded Carson its Morris Leibman Memorial Prize for his invention of single-sideband modulation, whose importance probably surpasses that of high-fidelity FM.[31]

Of course, the fact that Carson debunked only narrowband frequency modulation does not remove the chance that one way or another he inadvertently curbed further research in other kinds of FM, possibly by deterring practitioners who distorted his argument. Clarence Hansell, who himself often misunderstood Carson, credited "the ridicule heaped upon those who believed frequency modulation would permit reducing transmitter frequency bands" for the small number of articles about FM in the *Proceedings of the IRE* after 1922.[32] But patent records contrarily indicate that from 1922 to 1934 the number of frequency-modulation patent applications surged into the dozens, only four of which claimed to employ the narrowband method (see fig. 15 later in this chapter). The possibility that Carson discouraged research only in *narrowband* FM is more plausible, albeit admittedly harder to prove, given the small total of such patents ever issued. But blaming him for retarding the progress of frequency modulation generally is nonsensical.

FM Radiotelephony Research

Few false assumptions about frequency modulation have persisted more stubbornly than the belief that almost no one worked with the method until Armstrong invented wideband FM in 1933. Typically, assertions that John Carson's 1922 article essentially killed off FM research support this idea. Lessing declares that Armstrong "never lost an opportunity to rub it in that the investigation of frequency modulation had been fumbled."[33] But just as the traditional history of FM has twisted Carson's words, it has also completely overlooked the tremendous amount of useful work of dozens of people during the 1920s and early 1930s. In fact, approximately forty inventors filed for eighty-three patents directly and indirectly related to FM between the advent of the broadcasting boom in 1920 and the end of 1933, just before Armstrong revealed his first wideband FM system. The appendix provides a list of these patents.

The appendix requires some clarification. First, the modifying phrase FM-related in the appendix's title refers to patents that either explicitly mention the method or allude to frequency-modulation devices such as transmitters, receivers, and systems. Some patents also state that they describe only something that could be useful in frequency-modulation systems. Second, the appendix includes devices related to *phase* modulation as well. Without ignoring real differences between the two methods, this book presumes that nothing important distinguishes frequency modulation from phase modulation and that someone who has worked extensively with one of the methods has for all practical purposes worked with the other. Finally, although the appendix represents an attempt to list every FM-related invention filed before 1941, including patents with filing dates after 1940 proved impractical, because of the unmanageably large number of such inventions.

To place these inventions in context, this section provides an overview of FM research from 1913 to the mid-1930s, organized more or less according to the sources of FM-related patents, which roughly correlates with the significance of the inventions themselves: FM inventors of minor significance; the American Telephone and Telegraph Company; the Westinghouse Electric & Manufacturing Company (including its flagship broadcast station, KDKA); and RCA. Later chapters analyze in detail the work of Edwin Howard Armstrong, the inventor of "wideband" FM radio.

FM Inventions of Minor Significance

These nineteen patents comprise a wide variety of largely homegrown, sometimes eccentric, and never influential ideas. Before large corporations took up frequency-modulation research during the 1920s, nearly every FM patent in the United States was unassigned. A Grand Forks, North Dakota, man named Albert Taylor filed in 1919 an application for a simple radiotelephone patent.[34] In 1912 John Hayes Hammond Jr. applied for a patent for a dual-purpose arc oscillator system that employed FSK for radiotelegraphy and AM for radiotelephony.[35] Eighteen months later, Peter Cooper Hewitt came up with a peculiar frequency-modulated oscillator based on a heated chamber of vibrating gas or vapor.[36] Technically, Albert Van Tuyl Day filed his application for a complex vacuum-tube FM radiotelephony system in 1919, but the fact that his patent was issued thirteen years later diminishes the likelihood that the final document closely resembled his original idea.[37] Finally, in 1929 Hammond utilized frequency modulation in a communications system that transmitted light beams, not radio waves.[38]

The seven minor FM patents that their inventors assigned to companies other than Westinghouse, RCA, and AT&T before 1934 make only a slightly better impression than the unassigned inventions. General Electric and the German firm, Telefunken, accumulated two and four patents respectively. The Compagnie Générale de Télégraphie Sans Fil, of Paris, was assigned one patent, in 1931.[39] Patents from this group count individually as little more than curiosities, but they show that in Europe and America FM research did not lay dormant during the twenties.

FM Research at the American Telephone and Telegraph Company

Ten FM patents are assigned to AT&T and its subsidiary, Western Electric Company, all dated from 1920 to 1933. The Telephone Company undertook the job of making frequency modulation work with less coordination and energy than did

RCA and Westinghouse, partly because AT&T carried out its research in several locations. Also, the AT&T patents more often related to facsimile technology than to radiotelephony. By 1934 Western Electric employees had filed three radiotelephony and two facsimile applications, and AT&T engineers had applied for only one radiotelephone invention and three facsimile devices or systems. The tenth Telephone Company patent described a generic modulator circuit adaptable to the transmission of any kind of information.[40]

A combination of business and technological factors explains AT&T's relative lack of enthusiasm for FM radiotelephony. The corporation filed its last FM radiotelephony patent in 1926, the same year the firm permanently sold off its broadcast stations.[41] Furthermore, noise and fading presented much less serious challenges for *wire* telephone communications than for *wireless* radiotelephony. Finally, FM facsimile was easier to achieve than FM radiotelephony. As opposed to the real-time constraints of voice telephony, in which the reproduction of sound must occur in perfect synchronization with the source, no technical reason prevented the transmission of a still picture from taking minutes or even hours. Nonetheless, the fact that AT&T and Western Electric worked with FM at all weakens the assertion of historians of FM who have dismissed the 1920s as irrelevant.

FM Research at Westinghouse and KDKA

The first significant experiments ever with frequency-modulation radiotelephony began at Westinghouse's East Pittsburgh broadcast station, KDKA, during the early 1920s (see fig. 12 for KDKA's first FM patent). KDKA doubled as a research laboratory for Westinghouse, and the station's engineers, led by founder Frank Conrad, profitably pioneered several innovations in radio communications, most famously shortwave (frequencies above the upper limit of the standard AM broadcast band, at 1.5 megacycles) communications.[42] Only KDKA's historic local broadcasts of 1920 garnered more publicity than Conrad's trials of long-distance shortwave programs. From 1923 to 1925, listeners in the United Kingdom, South Africa, Australia, and north of the Arctic Circle could hear experimental transmissions that originated in Pittsburgh.[43] KDKA also led in the use of crystal-controlled oscillators for stabilizing transmitter carrier frequencies, which permitted more stations to broadcast with less risk of interference.[44] The station's research experience, particularly with crystal circuits, paid off during groundbreaking experiments with frequency modulation because FM demanded far steadier carrier frequencies than AM did. Westinghouse engineers also discovered that they

Aug. 10, 1926. **1,595,794**

D. G. LITTLE

WIRELESS TELEPHONE SYSTEM

Filed June 30, 1921

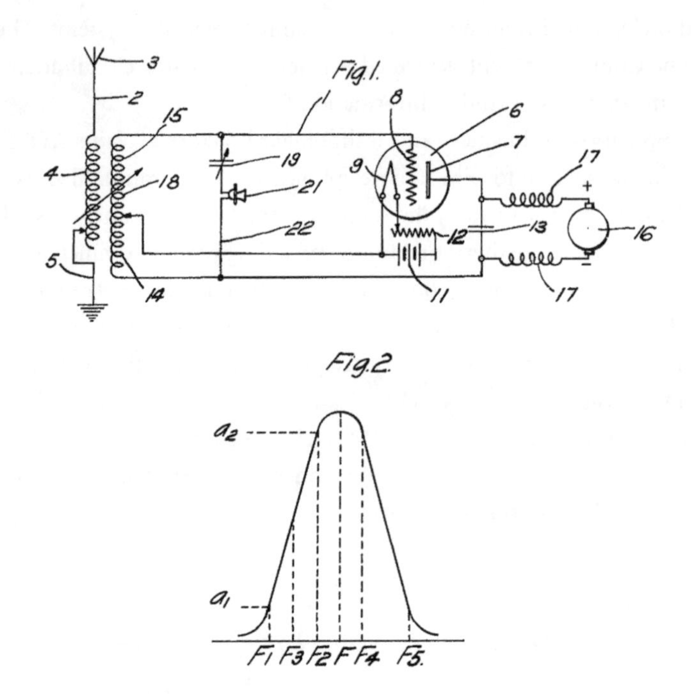

WITNESSES:

K. C. Clowes

H. L. Godfrey

INVENTOR

Donald G. Little

BY

ATTORNEY

Fig. 12. KDKA's First FM Patent, 1921. Upper figure shows a simple transmitter with a condenser microphone (21). Donald G. Little, "Wireless Telephone System," U.S. Patent No. 1,595,794, application date: 30 June 1921, issue date: 10 August 1926, assigned to Westinghouse. Little was an engineer for radio station KDKA during the 1920s.

could modify their crystal oscillators to make the first nonmechanical, electronic circuit for modulating a carrier's frequency.

Notably, the several inventors who held the FM radiotelephone patents assigned to Westinghouse lived near one another. Alexander Nyman, who in the summer of 1920 filed Westinghouse's first FM patent, for an arc-based radiotelephone, resided in Wilkinsburg, Pennsylvania. Virgil Trouant, the company's most prolific FM inventor in the period before 1934, lived in the same suburb. KDKA engineer Donald Little, of nearby Edgewood, filed the first all-vacuum-tube FM invention in June 1921.[45] Little's neighbor, boss, and friend, Frank Conrad, filed the station's second FM application seven months later (fig. 13). By 1934 these men, all Westinghouse employees and residents of metropolitan Pittsburgh, had nine FM patent applications to their credit.[46] For good reason this pattern of geographic proximity among FM's early inventors persisted into the 1940s: engineers accomplished more from discussing their work with colleagues than they did when working alone.

KDKA's FM research stemmed from the station's crystal-oscillator experiments, which began about the same time. Charles W. Horn, who eventually

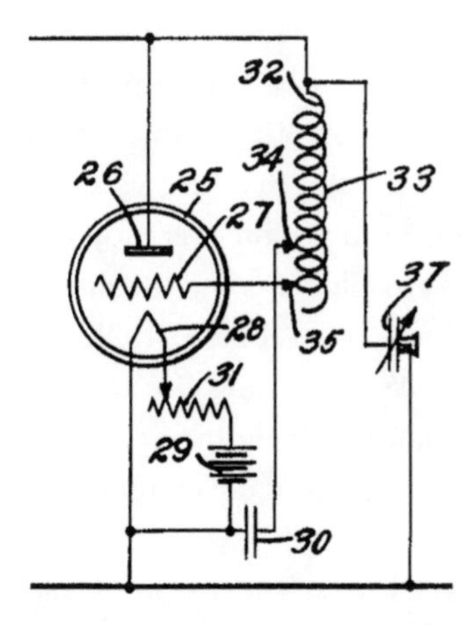

Fig. 13. Detail from Conrad FM Transmitter, 1922. Like Ehret's 1902 invention (see fig. 10), Conrad's device employed a mechanically coupled condenser microphone (37). Sound causes the capacitance of 37 to alter the electrical characteristics of a tuned *LC* circuit composed of condensers 37 and 30, and inductor 33. Consequently, the resonant frequency of the circuit varied with the amplitude of the sound. The significant improvement over Ehret's 1902 invention was Conrad's use of continuous waves and electronic amplification. Frank Conrad, "Wireless Telephone System," U.S. Patent No. 1,528,047, application date: 15 March 1922, issue date: 3 March 1925, assigned to Westinghouse.

transferred to RCA in 1929 and served as NBC's chief engineer, oversaw KDKA's engineering staff during this period. In 1939 Horn recalled that, while serving as KDKA's station manager, he and his engineers began FM experiments "soon after we first equipped KDKA with quartz crystal control" of the transmitter's main oscillator frequency.[47] Many radio stations adopted crystal-controlled oscillators for the purpose of improving frequency stability. But apparently only KDKA's engineers also noticed that an oscillator's resonant frequency shifts roughly in proportion to the voltage across its crystal, which led to the revelation that "we could vary the carrier wave frequency of the transmitter by changing voltage potentials across the crystal." Typically, an electrical audio wave was superimposed on the crystal, causing the crystal's voltage to vary, which in turn made the resonant frequency rise and fall in proportion to the instantaneous amplitude of the audio wave.[48] In fact, practitioners adopted this method as the most common means of frequency-modulating waves. Between 1926 and 1928, Westinghouse engineers filed five patent applications for inventions that used crystal-controlled frequency modulators.[49]

Examined as a group the Westinghouse patents reveal no explicit goals, but the fact that seven of nine described radiotelephone inventions suggests that the company briefly hoped to make frequency-modulation radiotelephony work in some useful way, most likely as a new kind of broadcast radio technology based on narrowband FM. Further, when Virgil Trouant filed Westinghouse's only narrowband patent in 1927, he revealed that "in one commercial embodiment of my invention, which has been in successful operation over an extended period, the . . . oscillator frequency is 970 kilocycles and the shift obtained with voice modulation is of the order of 800 cycles."[50] Indeed, KDKA often used 970 kilocycles to transmit its carrier. Trouant's description also corroborated Horn's recollection in 1939 that "we . . . were able to 'wiggle' the frequency of the [KDKA] transmitter over a range of some hundreds of cycles, the figure of 800 cycles was one of them."[51]

In 1938 the authoritative industry magazine *Broadcasting* stated that KDKA, "a number of years ago, conducted tests with frequency modulation but did not find them entirely satisfactory."[52] Several facts tell a more complicated story. A year later, Charles Horn made the broadest claims for what KDKA accomplished with frequency modulation during the 1920s. He implied, for instance, that his team discovered that FM suppresses noise before Armstrong did:

> We frequently discussed the need of some form of modulation, which was not dependent upon amplitude, in order that we might discriminate against amplitude modula-

> tion because we knew that static and noise were of that type of energy. In order to test this system, we made experiments with the receiver located in . . . the Machine Company Building at the E Pittsburgh plant, and which location was in about the noisiest place we could find due to the great amount of electrical machinery in that and neighboring buildings. I remember that the tests definitely proved that we could get a very much higher ratio of signal to noise when using the frequency modulation as against the amplitude modulation.

Horn added, "I received many complaints from listeners about their inability to hear KDKA during periods when we used frequency modulation," which "proved to me that it was true frequency modulation."[53] These claims, however, made years after Armstrong's invention of practical wideband FM, smack of wishful thinking, for only in a limited sense was Horn stating the facts. Ample evidence indicates that KDKA almost certainly made the first FM broadcasts, but no ordinary radio set could receive them. The complaints he cited proved only that listeners could not detect KDKA's transmissions. In fact, KDKA's FM technology never approached the requisite complexity that Armstrong's system achieved during the early 1930s, and no Westinghouse FM invention functioned on a practical level until the late 1930s. Most significantly, the station's engineers failed to grasp what Armstrong stumbled on years later: that static distorts the *amplitude* of a radio signal more than it distorts the *frequency,* a distinction that largely accounts for why wider FM channels make for quieter broadcasts.

That KDKA's experiments did not lead to something resembling the Armstrong system obscures their significance though, for in engineering one cannot infer worthlessness from a lack of practical success. Westinghouse's FM research did produce, in a roundabout way, useful results. The station's engineers must have learned from using 800 cps deviations what Carson taught with mathematical theory—namely, that narrowband FM did not work. Many of the rudimentary innovations KDKA's engineers devised appeared later in more sophisticated forms as components of the Armstrong system. Crystal oscillators, balanced amplifiers, frequency multipliers, limiter circuits, and high-efficiency nonlinear electronic amplifiers, for example, migrated into the material language that has characterized FM design since the 1920s. And though Virgil Trouant filed a single, poorly conceived patent application for a narrowband invention, he and other Westinghouse engineers correctly guessed that FM would eventually provide other advantages that the Armstrong system actually delivered: static reduction, resistance to fading, and greater transmitter efficiency. KDKA engineers, as one of the first groups to blaze the FM trail, made progress that seems unexceptional in a comparison that hindsight makes possible, and they pursued one dead end after another. But the hard work Westinghouse put into mapping out these dead ends later saved investigators at RCA—and Armstrong—from pursuing an uncountable number of false leads.

Knowledge of KDKA's work with FM spread, probably by word of mouth, to a handful of journalists and engineers far beyond Pittsburgh's environs. Oddly, despite evidence in the patent record that Westinghouse scarcely flirted with narrowband FM, for many years most published descriptions fixated on that mirage, suggesting that the central argument of John Carson's article had sunk less deeply into the minds of the larger community of radio practitioners than into those of the individuals who actually worked with frequency modulation. One can partly blame Westinghouse's managers for the confusion. In 1928 the firm's president, Harry P. Davis, disclosed in a speech at the Harvard Business School that KDKA had been "operating for some time with a different type of modulation called 'frequency modulation.'" He listed as one advantage the method's now well-known transmitter efficiency, which allowed the station "to eliminate three-quarters of the number of transmitting tubes that are required in the ordinary [AM] manner of transmitting. Further, the wave band is greatly sharpened and eliminates side band interference." Davis guessed correctly about the increased efficiency of FM transmitters, which unlike AM transmitters radiated constant-amplitude waves. But his use of the phrase "sharpened the wave band" and his implication that frequency modulation could achieve both spectrum conservation (by eliminating "side band interference") *and* noise reduction speaks to his ignorance of the proven futility of narrowband FM.[54]

The false myth of narrowband FM's potential also captivated a number of writers beyond Westinghouse. Mary Texanna Loomis's popular radio engineering textbook of 1928 stated that "a new kind of modulation, called 'frequency modulation,' has been used experimentally at KDKA." Like Harry Davis, Loomis hinted that FM promised spectrum conservation, declaring that "it is claimed for this [method] that stations can operate within a 10-kilocycle band." She cited FM's efficiency, too, and declared that FM stations "can also dispense with the usual [high-wattage] modulator tubes, with a reduction in power consumed."[55] KDKA's work probably also prompted Edgar Felix, an executive of the RCA-owned station WEAF in 1928 and a writer for *Radio Broadcast* magazine, to include narrowband FM among several innovations that might alleviate congestion. For FM, Felix wrote, "has been claimed the extraordinary virtue of accommodating simultaneously between one and two thousand broadcasting stations in the present band."[56] As late as 1930, similarly hopeful reports about narrowband FM appeared in Britain, where the respected magazine, *Wireless World*, published an article lamentably titled "Frequency Modulation: A Possible Cure for the Present Congestion of the Ether."[57]

That outsiders, however poorly informed, knew something of KDKA's FM work is unremarkable given the wide variety of unpublished media for the transmission of knowledge about radio technology. Much of the community of radio practitioners during the twenties exchanged information in ways that virtually ensured that word about the experiments would spread. Informality and camaraderie characterized the radio engineering profession, especially in northeastern states, where many practitioners who worked with FM lived and attended meetings of both the all-professional Institute of Radio Engineers and the "amateur at heart" radio clubs. And RCA's corporate structure, crafted to prevent Westinghouse, GE, and RCA from competing with each other, obviated any reason as to why Westinghouse engineers should not talk shop with colleagues employed by the other two companies.

The "Radio Group"

History tied the three corporations where FM research was most concentrated far more closely to each other than one might guess. David Noble describes how "the growth of the corporations, and the intensification of their control through trusts, holding companies, mergers and consolidations, and the community of interest created by intercorporate shareholding and interlocking directorates" characterized American business by the 1920s.[58] Noble's analysis makes a broad argument about corporate research in general, but it hits the mark with respect to FM radio. So does an examination of the companies that contributed most to frequency-modulation research during the twenties. Far and away the most important was the Radio Corporation of America. In 1919, with prodding by the federal government, General Electric created RCA primarily to keep American radio patents from falling into British hands. (GE itself had resulted from a merger of the Edison Electric Company and the Thomson-Houston Company in 1896.)[59] To minimize competition with its corporate owners, a contract banned RCA from manufacturing all but a small quantity of radio apparatus, but the company possessed the exclusive right to sell all of GE's radio products. RCA also acted as a clearinghouse that licensed radio patents to GE and other manufacturers on an equitable basis. Later, additional large companies that owned important radio patents joined RCA's "Radio Group" patent pool, which shielded from competition the large corporations that enrolled as members. These companies exchanged their patents and cash for RCA stock, and RCA licensed the same patents outside the Radio Group for profit.

By the time RCA began developing FM during the mid-1920s, the Radio Group had evolved into a close-knit, self-regulating syndicate. For the first few years, RCA's largest corporate shareholders included AT&T, American Marconi, the United Fruit Company, and Westinghouse.[60] Some of the group soon dropped out—most prominently AT&T, which sold off its RCA stock by mid-1923 and transferred the last of its radio stations to RCA's new network, NBC, three years later.[61] By the end of the 1920s, senior managers of only two corporate shareholders, Westinghouse and GE, dominated RCA's board of directors.

Two factors about RCA shaped FM research profoundly: first, the company operated a transoceanic point-to-point commercial service, a field no shareholder was permitted to enter; and, second, until the early 1930s, RCA, Westinghouse, and General Electric agreed not to compete with each other in the radio business.

FM Research at RCA

Despite the early lead in patents that Westinghouse built up during the early 1920s, RCA dominated FM research by the end of the decade. Engineers in the latter firm began investigating FM radiotelephony a few years after their KDKA colleagues did, but with different aims. While Westinghouse focused on making a commercial broadcast FM technology that utilized the medium frequencies of the standard AM band (500 to 1,500 kilocycles per second), RCA sought chiefly to improve the reliability of its point-to-point overseas shortwave service, which sent messages between stations separated by thousands of miles. Initially, the company's engineers evaluated FM from the standpoint of its effectiveness in solving the most critical problem of long-distance communications: fading.

Thanks to a large number of in-house memoranda by Clarence Hansell, who supervised engineering work at RCA's Riverhead, New York, laboratory during the 1920s and 1930s, we can learn a great deal about RCA's earliest FM work. In 1932 Hansell wrote that Harold O. Peterson and Harold H. Beverage had "set up and operated a simple frequency-modulation system in the Riverhead Receiving Laboratory" as early as 1924. Beginning in 1925, "the development of frequency modulated transmitting equipment suitable for high frequency experiments . . . had a regular place on the program of the Rocky Point [New York] Development Laboratory," and in 1925 RCA established experimental FM circuits between New York and stations located in Argentina and Brazil.[62] In 1927 Harold

Peterson strayed, for a short period, into the cul-de-sac of narrowband FM, when he applied for an FM radiotelephone patent, one of three assigned to RCA that year, and the only narrowband invention ever assigned to the company. "The requisite maximum wave band [i.e., the channel width] may be made as small as desired," explained his patent, which also claimed that a frequency swing "of only five hundred cycles per second . . . suffices for successful operation, even on very short wave lengths." Apparently, Peterson and his patent examiner had either not read, understood, or believed John Carson's 1922 article, for in contravention of Carson's argument, Peterson's patent also described the employment of a "frequency wobble which helps . . . avoid the use of side band frequencies, in the ordinary sense, with their attendant disadvantages."[63] (Carson, of course, had contrarily proved that narrowband FM creates innumerable side band frequencies.) In any case, FM soon after took a back seat to "other work," partly because of improvements in receiver antenna design that increased signal-to-noise ratios of AM radio signals. "Little concrete progress was made" with frequency modulation until 1929, when "interest in the problem was renewed and both the Rocky Point and the Riverhead laboratories began to follow it up more intensively."[64]

By mid-July 1929 the chief difference between the earliest patents of Westinghouse and RCA was that the latter company had filed a much greater number. RCA had accumulated some twenty-five FM patents versus Westinghouse's nine, a gap that would steadily widen until the mid-1930s. RCA's patents also claimed, more often than did Westinghouse's, to reduce the effects of fading, again a central problem for long-distance communications. But, in other ways, the two organizations did similar things; both turned out not radical designs but rather incremental improvements to FM, including circuits that Armstrong later further elaborated on and employed in his wideband system of 1933. The "limiter" circuit Clarence Hansell placed in the receivers of his August 1927 and October 1928 patents (see fig. 14 for one of the 1928 patents) to reduce the effects of fading, for example, appeared in a more sophisticated form in the Armstrong system.[65]

Patterns of FM Research

An examination of FM radiotelephone patent applications filed from 1913 through the 1930s indicates that the development of frequency-modulation radio occurred predominantly in three large corporations headquartered in the northeastern United States: RCA, Westinghouse, and, far less productively, AT&T. Aside from one-time independent inventors in Illinois, North Dakota, and the

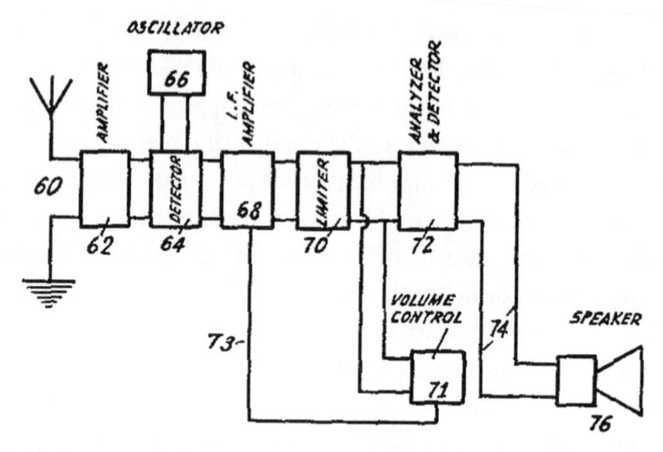

Fig. 14. Block diagram of Clarence Hansell's FM Receiver, 1928. Block diagram shows a limiter circuit, something Armstrong improved upon in his wideband FM system of 1933. Clarence W. Hansell, "Signaling," U.S. Patent No. 1,803,504, application date: 5 October 1928, issue date: 5 May 1931, assigned to RCA.

District of Columbia, and in a handful of European countries, the individuals who filed for patents lived in one of four adjacent northeastern states. New Jersey and New York together claimed more than two-thirds of the total patents; Massachusetts and Pennsylvania inventors contributed four and nine patents (5% and 11%) respectively. Where the inventors resided correlates with the location of the assignee corporations, because the largest companies that researched FM tended to employ the inventors who lived in these states.

Figure 15 and the appendix show how RCA came to dominate FM research for several years. That firm, whose engineers at the Riverhead and Rocky Point laboratories (both located less than eighty miles east of Manhattan) lived principally in the states of New York and New Jersey, acquired rights to half of the FM patents filed between 1920 and 1934 (forty-four out of eighty-three). In fact, 30 percent of all American FM patent applications before 1934 were filed by only two RCA research engineers, Murray Crosby (ten) and Clarence Hansell (fifteen), with Hansell alone holding nearly one-fifth of all FM patents filed.

Besides RCA, the other two major corporate assignees were AT&T and Westinghouse, each of which obtained the rights to ten FM patents before 1934. Nine of Westinghouse's were filed by men who lived near that firm's factory in Pittsburgh or in nearby Wilkinsburg and Edgewood. A weaker but nevertheless similar pattern existed for the ten patents assigned to the third major corporate source of FM patents, the AT&T–Bell Labs–Western Electric organization, although this

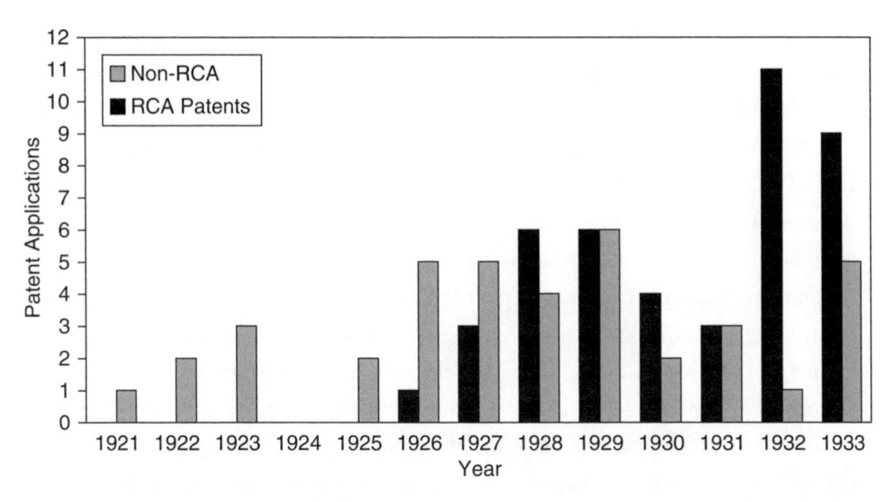

Fig. 15. FM-Related Patent Applications, 1921–1933. Darker bars = assigned to RCA; lighter bars = not assigned to RCA.

group's inventors were far more geographically dispersed among the three companies.

The chronology of RCA's patents suggests that interest in FM at RCA ebbed and flowed but, over the long run, persisted longer and remained stronger in that firm than in the other two major corporations. Figure 15 shows that RCA, which entered the field of FM later than AT&T and Westinghouse (filing in 1926 the sixth patent application of the early broadcast period), soon surpassed all other assignees combined. RCA engineers filed more than two-thirds of all FM patents in 1928, and in every year afterward through 1935 they continued to apply for at least half of the annual number of FM patents in the United States. But RCA's output occurred unevenly in three short-lived surges. The first began about 1926, peaked in 1928, and tailed off afterward. The second started in 1931 and peaked in 1932. The last surge began in 1934, shortly after Armstrong informed the firm about his wideband system. RCA's output the next year reached its peak with five patents, but in 1936 company engineers filed only one application, their last until the 1940s.

The smaller number of patents assigned to the other two major corporations makes discerning chronological patterns more difficult, but not impossible, to determine. Westinghouse accumulated nine FM inventions before 1934—only one-fifth of RCA's total—but the Pittsburgh firm began its research earlier, and sustained a continual trickle of patent applications from 1920 through 1928. Maximum output occurred during the two-year period from August 1926 through

May 1928, when five of ten Pittsburgh patents were filed. AT&T also owned ten patents, variously assigned to subsidiary firms Bell Laboratories and Western Electric (AT&T's manufacturing arm), as well as to the Telephone Company itself. All but one of these were filed in 1928 or earlier, only one before the publication of AT&T engineer John Carson's 1922 article. Even if readers widely misunderstood Carson's argument, one cannot argue that he prevented FM research from accelerating, even in his own company.

These facts support two broad generalizations about FM research before 1934. First, large organizations strongly shaped how FM radio technology evolved, confirming the argument of David Noble, who has contended that by the 1920s the age of the independent inventor had essentially ended, due partly to a concerted effort on the part of large corporations to co-opt independent research engineers and scientists. Second, the quantity of patent applications filed in the United States indicates the existence of at least a moderate level of curiosity about FM in a handful of research centers, well before Armstrong was issued his patents. True, no organization, not even the companies with the greatest interest, assigned the development of frequency modulation the highest priority. Several other fields of radio technology—antennas, vacuum tubes, linear circuits, and television—accumulated far more patents. But neither was FM an esoteric, neglected, or abandoned concept or practice during the 1920s, as Lessing would have us believe. (Indeed, the *Proceedings of the Radio Club of America* reported in 1939 that "what was probably the first public discussion of the subject was had before a meeting of the Club some fifteen years ago.")[66] All through the early broadcasting period, from 1920 through 1933, a significant number of men, many of whom were Radio Club members and engineers who knew Armstrong as a friend—corporate and independent alike—actively worked to develop practical FM technology.

During the 1920s, FM radiotelephony remained a solution searching for a problem. Because nothing more gravely challenged the future of radio during the 1920s than congestion, the mirage of narrowband FM opening up the spectrum for thousands of additional stations fascinated a few practitioners, even several years after John Carson demonstrated the method's infeasibility. FM seemed to offer other advantages as well, such as greater power efficiency. But, as the next chapter shows, it was fading, a problem of long-distance communications, that more than any other technical factor spurred FM research after 1929. Nevertheless, the work of the 1920s did not go to waste, and those who labored to make FM useful during the 1930s owed a great deal to the hits and misses of the previous decade.

RCA, Armstrong, and the Acceleration of FM Research, 1926–1933

> Major Armstrong feels that phase modulation and frequency modulation are extremely important developments and that we should keep it confidential for some time to come.
>
> *Harold Beverage, 26 February 1932*

The traditional history of FM radio implies that Edwin Howard Armstrong's revolutionary wideband FM patents caught RCA off guard. Lessing, for example, writes that "the saga [of FM radio] began shortly before Christmas, 1933, when Armstrong invited [RCA president] David Sarnoff up to the Columbia University laboratories to witness his latest wonder." Lessing says also that Armstrong offered to sell the patent rights to FM to RCA before all other firms because "R.C.A., by reason of the large royalties it had collected and the large research laboratories these had built up, was the logical company to undertake the expensive development of a new invention for the industry."[1] But the traditional history has it wrong. In fact, RCA and Armstrong both began researching frequency modulation about the same time—during the mid-1920s—and not surprisingly, because they enjoyed a collaborative relationship that made Armstrong, for all practical purposes, an RCA employee. His contractual obligation to RCA, not RCA's size, determined the "logical company" to develop wideband FM; Armstrong had no other choice.

The organizational context of the American radio industry strongly influenced the strategy of FM research before 1934 and the rate of progress in two other ways. From 1928 to 1933, RCA achieved far more with FM radio than did any other company, principally because a reorganization of the firm forced its largest corporate shareholders, Westinghouse and General Electric, to share their findings about FM with RCA engineers. Also, because RCA operated a long-distance communications service, its engineers saw fading—not static—as their chief problem. Unfortunately, the orthodox belief that a wider channel invariably passes more static noise discouraged these men from experimenting with wider frequency swings, an intellectual barrier that Armstrong would surmount in 1933 and 1934.

Armstrong's Relationship with RCA

The close relationship between RCA and Howard Armstrong grew out of Armstrong's boyhood experience as an amateur radio operator. He was born in the New York's Lower West Side neighborhood of Chelsea in 1890 to a schoolteacher mother and the American representative of the Oxford University Press. Armstrong joined the Radio Club of America, probably in 1911 or 1912, during his senior year as an electrical engineering student at Columbia University, and around the same time he made his first invention, the regenerative or "feedback" circuit. This truly revolutionary device made possible the electronic generation of radio-frequency waves, and it remains ubiquitous in electronic engineering today. When Armstrong demonstrated his invention to the American Marconi Company in December 1913, one of that firm's representatives was David Sarnoff, who soon became one of Armstrong's best friends.

America's entry into World War I advanced Armstrong's career considerably. He enlisted in the Army Signal Corps and served principally in France, where he co-invented the superheterodyne circuit, another fundamental radio invention that modern receivers still use to simplify tuning.[2] In recognition of this work, the French government decorated him with the Légion d'honneur, and the Institute of Radio Engineers awarded him its first Medal of Honor. Understandably, Armstrong took pride in his army experience, and for the rest of his life he preferred to be called "Major" Armstrong.[3]

After the war, his invention of the superheterodyne, regeneration, and a similar but never widely utilized idea called superregeneration, propelled Armstrong into the orbit of the Radio Corporation of America. In 1920 Westinghouse, one of RCA's major corporate shareholders, paid Armstrong $335,000 for the patent rights to the regeneration and superheterodyne patents.[4] This money, plus the hundreds of thousands of dollars he earned for selling additional patents, could have guaranteed his independence from corporations, but by the end of the decade, David Sarnoff, by now RCA's general manager, secured the inventor's complete loyalty with even more money. In June 1922 Sarnoff approved a payment to Armstrong of "$200,000 in cash and 60,000 shares of R.C.A. stock" for the superregeneration circuit, making the inventor, according to his biographer, Lawrence Lessing, "the largest individual stockholder in the company." The following summer, Sarnoff commissioned Armstrong and another engineer to design a home radio set, the Radiola, which "made more money for R.C.A. than any set that was to appear until 1927." For this Armstrong collected an additional 20,000 shares of stock. By the winter of 1922–23, Armstrong's holdings in RCA amounted in value to "something over $3 million," a figure that tripled by 1930. Because he had also signed an agreement promising to grant RCA first refusal on his future inventions, for all practical purposes he was a consulting engineer with only RCA as a client. Thus, for RCA to share proprietary information with Armstrong was to benefit both him and the firm.[5]

Armstrong knew during the 1920s that RCA was working on FM; indeed, that knowledge likely inspired him to pursue his own line of research. Harold Beverage, who recalled first meeting Armstrong at a Radio Club of America gathering "soon after he'd come back from France," counted the inventor as one of his "close friends" and visited Armstrong's penthouse apartment "quite often." In 1992 Beverage remembered that Armstrong first heard of RCA's frequency-modulation research during the 1920s when "a guy named Murray G. Crosby was working on all kinds of modulations [at RCA]: amplitude, phase, frequency and any other ones you can cook up. Armstrong was interested in that." And, as Beverage explained, "Armstrong was free to come to Riverhead and see what we were doing. And he did. He came out quite frequently."[6] He even got on well with the company's lawyers, once receiving a letter of thanks from an RCA attorney for volunteering to defer to RCA in a patent interference case.[7] In 1931 the company's Patent Department sent him a list of "patents [that] may prove of interest to you in connection with your work on frequency modulation."[8]

Armstrong never kept good records of his work, but ample evidence confirms that his research with frequency modulation began shortly after RCA's. In 1927 he filed a patent application for his first FM radiotelephony invention.[9] Two months later Harold Peterson, who worked closely with Beverage, filed RCA's first in-house FM application.[10] The fact that these two patents constitute half of all the narrowband patents ever issued by the United States Patent Office suggests that the men who filed them were cognizant of each other's work.

Unification, Convergence, and the Acceleration of FM Research at RCA

Historian Hugh Aitken has observed that the companies that owned RCA during the 1920s—namely, GE, Westinghouse, and AT&T—normally did not share engineering data, largely because logistical barriers discouraged the movement of notebooks and other records from one office and field laboratory to another.[11] But Aitken limits his analysis to a period that terminated around 1928, when RCA verged on a series of organizational changes that for FM radio dislodged these obstacles. Those changes were summed up in one word: "unification." Even before 1928, David Sarnoff, the general manager of RCA, and the person who most clearly envisioned the advantages of centralized research, saw waste in the uncoordinated distribution of radio manufacturing and sales functions among GE, RCA, and Westinghouse. He thus launched a campaign to consolidate all operations related to radio. The idea made sense commercially, because the three firms were already contractually barred from competing with each other in almost all aspects of the radio business. Unification would also reinvigorate frequency-modulation research.

In October 1927 representatives of Westinghouse, GE, and RCA appointed Sarnoff chair of a three-man committee to evaluate his own plan. Six months later, the panel recommended creating a new "Radio Manufacturing Corporation," with three shareholders: GE (48%), Westinghouse (32%), and RCA (20%). With this goal in mind, RCA's board of directors, many of whom also served GE and Westinghouse as top-ranking executives, approved a leveraged purchase of the Victor Talking Machine Company for nearly $70 million, a move that not only gained RCA entry into the lucrative phonograph market but also enabled the company to mass-produce radio apparatus. On 26 December 1929 the RCA Victor Company was incorporated, and on 3 January 1930 the board of directors made Sarnoff president of RCA.

Historians have characterized unification as a grave setback for Sarnoff and RCA because an ensuing federal antitrust lawsuit shut down the reorganization project during the early thirties, which ultimately led to the breakup of the Radio Group patent pool and its noncompetition agreements.[12] But unification accelerated the progress of FM radiotelephony as much as any single cause by creating a temporary intellectual pipeline though which knowledge acquired about FM flowed directly from General Electric and Westinghouse to RCA, and eventually to Howard Armstrong.

New information about FM began to arrive at RCA laboratories several months before the formal acquisition of the Victor Company. In July 1929 James Harbord, the future chairman of the board of RCA Victor, assigned C. H. Taylor, the new vice president of engineering at RCA Communications, the job of merging the research efforts of Westinghouse, GE, and RCA. "I consider this coordination of the engineers of the three companies," Harbord wrote to Taylor, "as being one of your major duties." Harbord instructed "the Westinghouse Company . . . to supply a report [to RCA] on the frequency modulation method of broadcasting done [by KDKA] at East Pittsburgh." FM, Harbord believed, represented one of the technologies "which are vital to the success of our new company." Without "coordination," the efforts of the three companies acting independently, he feared, "will lack direction, cohesion and results, besides being expensive."[13]

As the unification plan fell into place, RCA began two FM projects, both built on a foundation that Westinghouse and KDKA had already laid. The first endeavor began around 1928 or 1929, when the National Broadcasting Company (NBC), then owned by RCA, installed Westinghouse-built receivers in the network's Manhattan-skyscraper headquarters. Engineers hoped to detect transmissions from KDKA, but the experiment swiftly came to naught because the transmitters and receivers—prototypes that a team led by Frank Conrad had designed—continually drifted out of electrical alignment. After one month of tests, NBC engineer Robert Shelby reported a raft of equipment failures, and he speculated that the Westinghouse apparatus was "not properly constructed, or is not properly adjusted, or the transmitter is not properly adjusted to work with frequency modulation."[14] Soon afterward, mention of this project disappears from RCA company records.

RCA Communications' second undertaking achieved far more, thanks principally to the funneling of data about FM from GE and Westinghouse to RCA. Ironically, RCA researchers learned from reading GE's laboratory reports that General Electric had almost nothing to teach about frequency modulation. J. L. Labus, the principal author of one paper, did little but summarize information he had gleaned from the scant published technical literature about FM and from his recent visits to the RCA laboratories in Rocky Point and Riverhead. Not surprisingly, RCA's best FM theorist, Murray Crosby, found little admirable in, as he saw it, Labus's error-riddled paper. Crosby surmised that Labus's expertise was "confined to the mathematics side of the question and that his practical experience and knowledge are very meagre. He tries to cover his uncertainty with borrowed (from R.C.A.) experiences and fails in the attempt. Perhaps if he confined his writings to the mathematics he would not run into this trouble." But Crosby also pointed out several mathematical errors in the report, as well as a misinterpretation of something Labus observed during a visit to Rocky Point. Labus claimed that he had learned from RCA that a frequency shift "must not exceed 500 cycles in order to produce a quality comparable with that obtained in amplitude modulation." In fact, Crosby explained, Labus had inadvertently based his analysis on a broken transmitter. The malfunction prevented the transmitter from swinging more than 500 cps from the center frequency, and Crosby guessed that Labus had wrongly inferred that the transmitter's behavior—suggestive of narrowband FM—was normal operation.[15]

Westinghouse engineers, in spite of their inability to design a functional system, made a far more positive impression, primarily because some of the Pittsburgh engineers had begun to tease out a respectable mathematical theory to describe FM. In September 1929, a few weeks after Harbord issued his unification edict, Clarence Hansell read a report written by V. D. Landon, who at the time held the position of chief operator of KDKA.[16] Before unification, Landon had intended to submit his manuscript to the *Proceedings of the Institute of Radio Engineers,* and Hansell agreed that his new colleague's article "certainly should be published." But not immediately, he cautioned his supervisor, lest the piece motivate competitors outside the Radio Group also to take up frequency modulation. "It seems very probable," said Hansell, "that the publication of Mr. Landon's paper would advertise the possibilities of frequency modulation very widely and cause many others to work on this same development." Better, he suggested, to wait until RCAC gains a surer foothold in the area of long-distance FM. "Present indications are that this method may become an exceedingly important one," explained Hansell, adding that "we believe that no one except ourselves is following up the development of frequency modulation." Therefore, he recommended, "RCA [should] build up a very strong patent situation by intensive development before rival companies become interested in the method."[17]

One hitch prevented RCAC from fully exploiting the Westinghouse data: few engineers at Riverhead could understand Landon's analysis. To overcome this problem, the job of translating his findings, along with those of other theoretical literature, into more accessible language fell on Murray Crosby. In June 1930 he distributed a memorandum to RCA managers and engineers that explained the nuances of frequency modulation in terms that less theoretically inclined engineers could understand.[18] As a grateful Harold Beverage confessed at the time, Crosby "convincingly [cleared] up many points which were obscure to engineers, like myself," and RCA's vice president of engineering agreed that the report gave him "a clear picture of the phenomena involved" with frequency modulation.[19] Every engineer involved with FM at Riverhead and Rocky Point likely received a copy of Crosby's report, including Howard Armstrong, a frequent visitor.

Deeper theoretical knowledge, practical experience, and a surer picture of how much—and how little—Westinghouse and GE had accomplished stiffened RCA's resolve to make FM work. On 17 October 1930 Hansell asked Ralph Beal, the manager of the Pacific Division of RCAC's overseas service, to assist in a long-term FM experiment aimed at improving long-distance telephone and telegraph communications. Beal, whose responsibilities included overseeing a former Marconi Company station in Bolinas, California, located near San Francisco, expressed "a great deal of interest" in Hansell's proposal and suggested daily four-hour tests between the West and East Coasts, on shortwave frequencies of around 9, 14, and 18 megacycles.[20] Hansell agreed, and after two months of preparing prototype apparatus, the Riverhead and Rocky Point engineers were ready to begin. Hansell and Beverage assigned the job of running the California side of the trials to a junior engineer named James Conklin, who set out by automobile from New York for the West Coast at the end of January 1931. By mid-March, Conklin had installed several FM transmitters and receivers in an old storage building located at the Bolinas station, and from mid-April until September, he helped conduct the most ambitious experiments with FM to date. In the beginning no one could confidently predict success, but an optimistic Harold Beverage declared beforehand that "we have some reason to hope for a worth while improvement in . . . short wave communication."[21]

Conklin's homesickness accounts for much of what we can know about these experiments. Aside from tinkering with the equipment, he had little to relieve the tedium of recording measurements of repetitive test signals from New York and sending similar patterns back east. Moreover, his colleagues in Riverhead initially shared frustratingly little information about the results of the trials. In June he asked for permission to return to New York. Hansell turned Conklin down but shored up his morale with a lengthy explanation of what, so far, had been achieved. "[Your] inability to obtain satisfactory replies to requests for information as to what conclusions are being drawn from the tests," said Hansell, ". . . may have given you the impression that interest in the tests is somewhat lukewarm on this end of the circuit." In fact, the experiments had produced both positive and negative results, and hence "the outcome is still in doubt." Hansell also disclosed that Howard Armstrong was trying out a new receiver of his own design. "Your transmissions," Hansell assured Conklin, "have been observed on the equipment developed by Crosby at Riverhead and also on equipment developed by Major E. H. Armstrong."[22]

It is important to understand that a long-distance commercial system already in place, rather than a short-range high-fidelity broadcast service not yet imagined, shaped how RCA designed and tested FM at this stage of the technology's development. For more than a decade, the firm had been in the business of sending and receiving long-distance point-to-point radiotelegraph and radiotelephone messages. The Rocky Point and Riverhead engineers therefore hoped to adapt FM for that kind of work, primarily to reduce the effects of fading. Of merely secondary importance was the expectation that FM transmitters might cost less to build and operate. And no one at the time, including Armstrong, seriously envisaged expanding the scope of FM to encompass other uses, such as commercial broadcasting or short-range radio communications of any kind. Even more removed from consideration were dreams of what later FM pioneers called "staticless" radio or of a high-fidelity medium.

Ultimately, the Bolinas tests revealed long-distance FM to be neither a simple failure nor an unqualified success. On occasion, frequency modulation worked well, sometimes dramatically better than AM, but only during the beginning and end of daylight hours. For the remainder of a twenty-four-hour day, Hansell told Conklin, "it was obvious . . . that the quality of reproduction obtained with frequency modulation was considerably inferior to the quality obtained with amplitude modulation." FM, Hansell explained, faltered most often "when transmission conditions were such as to allow reception of several incoming rays of radiation."[23] In other words, frequency modulation fell victim to an old problem in long-distance radio communications: multipath fading. The Bolinas transmitter simultaneously radiated multiple waves that ricocheted between the ionosphere and the earth's surface from California to New York. Because each wave traveled a unique and unpredictable path, two waves that left a transmitter at the same instant might arrive out of phase because one traveled a longer path. If a pair of waves traveling from Bolinas arrived exactly 180 degrees out of phase at the receiving antenna in New York, the net effect at the receiver would be a cancellation of both waves. Sporadic multipath fading also degraded the reception of AM-modulated waves at almost all frequencies, but RCAC discovered that FM in the ultra-high frequencies was even more susceptible.

To their credit, Hansell and Crosby made no final judgment about frequency modulation in general, particularly after Armstrong dropped in on the Riverhead laboratory on 19 June 1931 to listen to transmission tests from Bolinas. He announced that while eavesdropping on their tests "in the middle of New York City" he had gained "a much more favorable impression of the possibilities of frequency modulation than had been obtained by the observations at Riverhead." Armstrong invited his colleagues to hear for themselves, and so, six days later, Clarence Hansell, Murray Crosby, and Harold Beverage motored the seventy miles from Riverhead to Columbia University "to observe the reception and to discuss plans for future tests."[24]

The evening of 25 June 1931 ranks among the most significant in the history of frequency modulation, thanks in large part to the considerable talents of Armstrong's guests. Within a few years Murray Crosby would begin writing articles and filing patents that established his reputation as one of the twentieth century's pioneering figures in the field of frequency-modulation theory. Clarence Hansell had already applied for a dozen FM patents—one fourth of all that had ever been filed—and had been issued three. Though the FM patent record of Harold Beverage would never rival Crosby's or Hansell's, he had nonetheless made a name in the field of antenna design. Further, few men, even at RCAC, had more extensive hands-on experience with frequency modulation systems than Beverage did. By comparison, Armstrong was a latecomer; despite his stellar reputation in other areas of radio engineering, he had in fact no FM patents to his credit; and one of the two he had applied for described an unworkable narrowband invention.

Crosby and Hansell already knew a little about Armstrong's setup, as Armstrong had described it to them during his recent visit. Upon arriving in New York City, Hansell "immediately observed" that reception conditions were "very much different" from those in Riverhead. (Presumably he heard local transmissions from Riverhead, and not from Bolinas, whose signals multipath distortion usually rendered inaudible on summer nights.) In the city, "inductive disturbances"—that is, manmade static—from "all sorts of electrical equipment, buses and motor cars" "almost completely drowned out" AM reception. In contrast, "the reception of frequency modulated signals was often reasonably good and in practically all cases was much better than the amplitude modulated signals." "It is safe to say," estimated Hansell, "that the signal to noise ratio for frequency modulated reception was at least 10 to 1 in voltage better than amplitude modulation." He hoped that directional antennas might alleviate the relatively poor performance of AM but admitted that on balance "it seems probable that the results obtained might be overwhelmingly in favor of frequency modulation."[25] These tentative statements made up some of the first reports of FM's most famous property today: its ability to suppress static, particularly the inductive static that originates most often in urban areas.

Commercial reality dampened Hansell's enthusiasm, though. To be sure, frequency modulation, even in its pre-high-fidelity stage of development, resisted, far better than AM, disturbances caused by static. Within five years, broadcast FM pioneers outside RCA would justifiably cite this feature when they promoted Armstrong's new "static-less radio." But RCAC's chief business was to provide point-to-point communications between distant stations, which seemingly ruled out long-distance FM because of the method's susceptibility to multipath fading. Moreover, static suppression counted less as an advantage for RCA than one might assume. Because the company preferred to locate stations on sites with minimal electrical interference, it chose locations far removed from urban areas like New York City. No one before the Bolinas tests would have thought about comparing reception in rural Riverhead with reception in the electromagnetic din of Manhattan. As Hansell put it, "in our [RCA] Communications system it is not customary to do the receiving in cities." Thus, he concluded, "prospects for the general adoption of frequency modulation . . . for long distance telephony and multiplex telegraphy are not very good."[26]

Nevertheless, what Armstrong had accomplished so fascinated his guests that they lingered until three o'clock in the morning. Armstrong's showpiece was a fifteen-foot "breadboard" bench, on which he had wired a prototype system. The transmitter impressed Crosby and Hansell with circuits that boasted greatly improved linearity and lower distortion, two properties upon which FM has always depended. Hansell and Crosby suggested modifications to match RCAC's mission of providing "good enough" radio telecommunications; they wanted to simplify the circuits somewhat without forfeiting the long-dreamed-of advantages of long-distance FM—namely, its "ability to eliminate fading" and its efficiency, that is, "the possibility of obtaining four times the transmitter power output that can be obtained with amplitude modulation without appreciably changing the cost of the transmitter." Hansell conceded that his and Crosby's proposals would allow for more noise than did Armstrong's original design, but the results would be no worse than AM delivered under optimal conditions. "Crosby and I concluded," he explained to Conklin, "that we should obtain all the advantages of frequency modulation [with our changes] except for some sacrifice in signal to noise ratio and at the same time we should obtain a quality of reproduction substantially the same as for amplitude modulation."[27]

Clarence Hansell recorded no reaction on Armstrong's part to these suggestions, but a few days later Hansell wrote to Conklin that the inventor had asked RCAC to ship his breadboarded transmitter and receiver to Bolinas. Despite his reservations about using FM for long-distance work, and about the risk of transporting a fragile prototype thousands of miles, Hansell almost approved Armstrong's request. FM had long been a sideline interest of Hansell's, and he predicted to Conklin that "the present development [of FM] will prove to be an extremely important one and may result in a great improvement in short wave telephony and multiplex telegraphy." He also began to imagine other uses besides long-range communications, adding that frequency modulation "may also be the forerunner of a solution to the problem of very short wave broadcasting, television, etc. within city areas where the induction problem is the biggest obstacle to satisfactory service."[28] This prediction did not describe exactly the high-fidelity system that Armstrong would patent eighteen months later, but it was close, and after all, Hansell's work had little to do with broadcasting. Besides, Armstrong saw the future of FM no more accurately than Hansell did.

Sizing up what RCAC had done with frequency modulation by the end of 1931 depends on one's yardstick. Hindsight tells us that the company's engineers were pursuing a futile game by attempting to perfect long-range FM radio. Moreover, the belief that a wider frequency swing let in more static especially impeded progress. In 1931 RCAC engineers were working with swings no greater than approximately 25 or 30 kilocycles, while the Armstrong wideband system of two years later employed a swing of 150 kilocycles. No one in 1931—not even Howard Armstrong—hoped, let alone predicted, that further widening the swing and shortening the distance between transmitters and receivers would bring spectacular benefits—namely the suppression of static to nearly inaudible levels, the tripling of the audio bandwidth, and the consequent improvements in audio fidelity. In fact, RCAC engineers were technologically and psychologically committed to the goal of *minimizing* the frequency swings—not below, of course, Carson's well-known theoretical limit of twice the audio bandwidth, but neither did company engineers wish to widen the swing needlessly and thus waste spectrum. Furthermore, Clarence Hansell intuited that a wider swing, which created additional sideband components, would incur greater audio distortion, especially at lower carrier frequencies and longer distances. He correctly theorized that because power distributed among sideband components over a wider channel would be spread more thinly, then weaker components would more likely risk attenuation to inaudible levels. Again, Hansell's disinclination against wider swings arose from RCAC's mission of running a long-distance service; he had little incentive to contemplate what might happen to those sidebands over short ranges.

In terms of their own goals for FM, RCAC engineers had good reason to anticipate further progress. With Armstrong's help, they had come closer than anyone before to realizing a practical technology with an audio fidelity comparable to the quietest AM system. Building on Westinghouse's work, RCAC had discovered empirically much about what FM could and could not do. Most important, the oldest dream of FM experimenters, implied in Cornelius Ehret's 1902 patent, had materialized: FM consistently resisted fading, albeit at relatively short distances. FM transmitters also demonstrably wasted less power than did AM transmitters, proving an intuitive theory first spun by Westinghouse engineers. The location of Armstrong's receiver suggested that FM could sometimes be received more clearly than AM in electromagnetically noisy environments—that is, it resisted static. Finally, they had learned where FM failed. Narrowband FM did not work at all, and the Bolinas tests indicated that multipath distortion would always plague long-distance FM.

All in all, only an institutional predisposition against experimenting with wide channels and short distances prevented RCAC from developing a commercially practical FM broadcast system, a prejudice manifested by Hansell's distaste for wider swings. But no reason existed to assume that this bias would have been permanent or that RCAC's top engineers had not the ability to devise wide-swing circuits. Even without Armstrong's help, RCA almost certainly would have achieved some kind of practical FM, though that system would likely have lacked the high-fidelity features of the Armstrong version that lay ahead.

Despite the well-founded skepticism surrounding FM's future for RCAC's long-distance work, on one occasion the technology worked fabulously for that very purpose. Eight days after Hansell, Beverage, and Crosby met in Armstrong's lab, an NBC executive in New York telephoned Ralph Beal in Bolinas to request that the California crew relay from Cleveland, Ohio, to Hawaii and the Philippines the following evening's heavyweight boxing championship fight in Cleveland between Germany's Max Schmeling and Young Stribling of Georgia. NBC routinely relayed broadcasts to Asia, accomplished with a telephone circuit linking Bolinas to an inland source like Cleveland, then with a point-to-point shortwave AM radio circuit from Bolinas to stations in Hawaii and the Philippines, which rebroadcast locally on standard AM frequencies. James Conklin, who happened to be visiting Beal during the telephone call, suggested taking advantage of the event to compare shortwave transoceanic AM with shortwave transoceanic FM—that is, to relay the telephone feed from Ohio to Bolinas, and then to the overseas stations, using separate channels for each method. Beal assented, recalling that fading and interference during the previous night had wiped out an amplitude-modulated music program beamed to Manila.

As the bout approached, Conklin and Beal realized that if FM could not carry the fight to the Philippines, nothing would, for noise on AM receivers in Manila had climbed to intolerable levels. After Conklin made a few adjustments in his equipment, for the first time FM radio began to air a complete program. What followed was a spectacle to which overseas listeners, none of whom knew about FM's role, responded enthusiastically. AM Station KGU in Honolulu reported 140 telephone calls from fans who, after Schmeling TKO'd Stribling in the fifteenth round, complimented the unusually good reception. The operator of RCAC's Manila station noted in a telegram to Conklin a low audio-frequency distortion similar to what Hansell had also observed in Armstrong's lab, but the same operator pronounced the experiment "wonderful." "MANY THANKS," he telegraphed Conklin, "FIGHT BROADCAST SURELY GREAT THING FOR ORIENT."[29] Beal marveled in his follow-up report that "I doubt very much if Manila would have received anything worth while had the X [frequency] modulation not been available. If these results are duplicated in the future," he predicted, "there can be no question as to the merits of this type of modulation for long distance work."[30] When Hansell forwarded to Armstrong a copy of the Manila operator's telegram, Armstrong, whose equipment had *not* been used, groused that it was "too bad they didn't have the right receiver."[31]

The working relationship between RCA and Armstrong was complicated. On one hand, their interests were so legally and financially intermingled as to make the firm, for all practical purposes, Armstrong's exclusive client and Armstrong almost an RCA employee. On the other hand, it would be going too far to describe Armstrong and RCAC as collaborators in the usual meaning of the word, because the company's engineers revealed far more to him about their day-to-day work than he disclosed to them about his. (This lack of reciprocation, though, never troubled RCA engineers, who gave Armstrong wide berth because he was obligated to offer RCA first refusal for his patents.) Nor can one claim, without qualification, that RCAC engineers co-invented the Armstrong system. But clearly Armstrong owed much to his friends at Riverhead, because his insider status gained for him knowledge about the successes and instructive failures of Hansell, Beverage, Crosby, Conklin, and other Riverhead engineers—knowledge he applied to his own work. As figure 16 shows, Clarence Hansell and Murray Crosby forwarded to him detailed descriptions of test results, proposed circuit changes, and other matters relating to FM.[32] Armstrong also knew that RCAC engineers filed eighteen patent applications for FM inventions during the trials, including six by Crosby, three by Hansell, and two by Conklin. He himself filed two, in January 1933, both for his wideband system, but he concealed this fact until close to their issue date in late December of the same year.

Rocky Point, New York
July 16, 1931
A-RD-1105

Major Edwin H. Armstrong
211 Central Park West
New York, New York

Dear Mr. Armstrong:

I believe Mr. Beverage advised you by telephone that I would send a letter outlining some of my conclusions in connection with the study of phase modulation and of equipment with which to carry it out. I have not been able to give the subject nearly as much time as it probably deserves so that my conclusions are not complete and have not been gone over thoroughly to detect errors. However, by attempting to state them to you I may be able to learn something myself and, at the same time, provoke discussion which may result in a better understanding of the problem by all of us.

First of all it appears that, if we are to obtain good quality reproduction over long distances, the number of side frequencies produced by the modulation must be kept to a minimum. To do this we must keep the phase deviation of the transmitted current to a value much less than plus and minus 90 degrees. I believe that plus and minus 45 degrees phase deviation in the final output of the transmitter should be the maximum allowed for peak values of the modulation current. This corresponds to side frequencies having a maximum voltage sum equal to the carrier and is substantially similar to 100% amplitude modulation.

Since the carrier power of a phase modulation transmitter may be made four times the power of an amplitude modulation transmitter, with equal number of tubes and about the same cost, it is evident that we will obtain approximately twice the carrier and side band currents if we use phase modulation. Since the detector efficiencies are equal in receivers correctly designed for either method, it appears that the strength of receiver output, where no limiting is used, will be four times as great for 100% phase modulation as for 100% amplitude modulation.

At the same time, if the two filters for the detectors of the phase modulation receiver have characteristics such as shown in Fig. 1, only one sideband reaches each detector. Therefore the

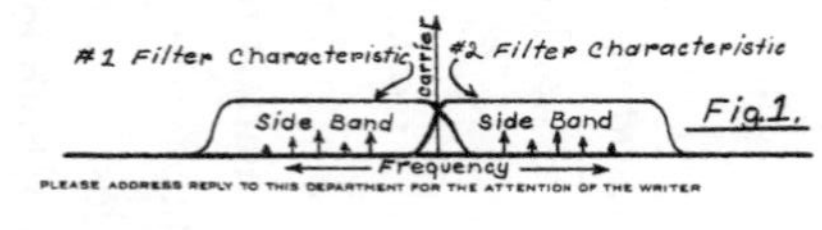

PLEASE ADDRESS REPLY TO THIS DEPARTMENT FOR THE ATTENTION OF THE WRITER

-2-

ratio of carrier to side bands in each detector is twice as great as is the case in an amplitude modulation detector. This should result in less harmonic distortion and cross modulation products from beating between side frequencies.

Another interesting observation is that, in a phase modulation receiver with the characteristics illustrated in Fig. 1, it is not necessary for the carrier to remain stronger than the combined peak value of the sidebands. The terrible distortion observed with amplitude modulation when the carrier fades below the sidebands should therefore be absent.

Although on first inspection the use of filter circuits such as those shown in Fig. 1 may seem as difficult as the reception of short wave single sideband transmission I do not believe this to be the case, because the carrier frequency is automatically fixed at the transmitter. By applying automatic control to the first beating oscillator in the receiver, operated by detector current unbalances, as you have done in your receiver, the problem of keeping the receiver adjusted should not be difficult.

On the assumption that the final phase deviation in the output of the transmitter should be kept below plus and minus 45 degrees, I have given consideration to the relative merits of the transmitter set up in your laboratory as compared with the one which I suggested to Conklin in my letter of June 27, 1931, a copy of which was sent to you.

In the output of your modulator appears the carrier, which may be as indicated in Fig. 2. To this carrier are added two side frequencies, or groups of frequencies, which may be represented as two sets of vectors, one rotating faster than the carrier and one slower. In other words, one rotates counterclockwise and one clockwise with respect to the carrier. They have a phase relative to the

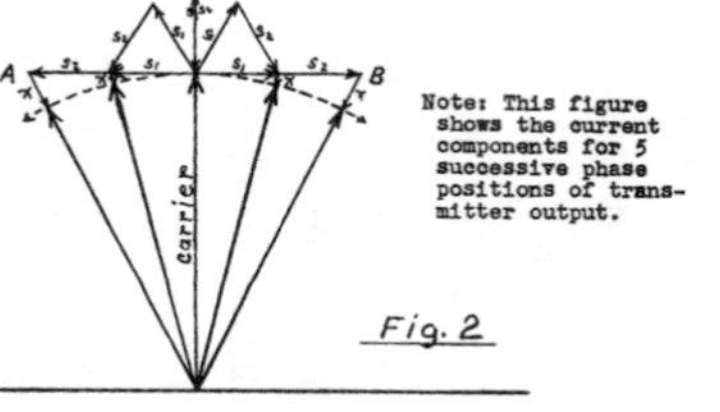

carrier which is displaced 90 degrees with respect to the phase relations obtained with amplitude modulation. Therefore, the sum of any pair, or any number of pairs of side frequencies must always be at right angles to the carrier and will lie on the line A-B (Fig. 2). The peak radio frequency output from your modulator is then equal to the vector sum of the carrier and the peak value of the sidebands. The output power from the modulator is increased by the sidebands.

As soon as the output power is made constant by limiting, the carrier is reduced and additional side frequencies are produced. These added side frequencies correspond to the addition of both amplitude and phase modulation components. Their vector sums are indicated as X in Fig. 2. It will be apparent that these undesired components are small for small phase deviations but become extremely large when 90 degree phase deviation is approached.

Since this addition of undesired frequency components is determined by the phase deviation irrespective of the carrier frequency and the deviation is multiplied by multiplying the carrier frequency, it is apparent that the phase deviation in your modulator should never be more than about plus and minus 45 degrees divided by the multiplying ratio of the frequency multipliers.

I have not been able to devise any modulator which would be free from the undesired frequencies produced by limiting without sacrificing the high power efficiency and output from a transmitter and have concluded that the only practical solution is to keep the phase deviation low. Consequently, I believe that, from the standpoint of theoretical performance at least, your modulator cannot be improved.

In considering the extent to which the relatively simple form of modulator suggested to Conklin meets the performance of your modulator, I have constructed the vector diagram shown in Fig. 3. In this diagram C_1 and C_2 represent the output from the two tubes used in the modulator, the vector sum of which gives the carrier. The relative values of C_1 and C_2 are varied differentially by the

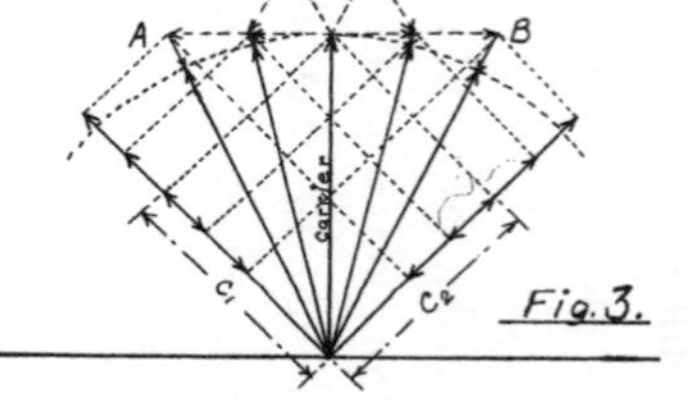

Fig. 16. Part of Hansell Letter to Armstrong about FM, 16 July 1931. Three pages of a letter written by Clarence Hansell to Howard Armstrong, discussing RCAC's FM research and development. Drawing on leftmost page indicates Hansell's theories about sidebands created by frequency modulation. The other drawings depict Hansell's vector analysis of frequency modulation. Hansell to Armstrong, 16 July 1931, box 161, AP.

Armstrong also persuaded RCA to adopt of a policy of nearly complete secrecy about frequency modulation, which explains why the company adopted the term *X modulation.* "Major Armstrong is also following your tests with considerable interest," Beverage told Conklin's replacement in California. He added that "the main reason why we do not wish to refer to the word 'modulation' and its qualifying adjective ['frequency'], is, that we do not wish to let outsiders, who may be listening in, learn what we are doing." Moreover, "Major Armstrong feels that phase modulation and frequency modulation are extremely important developments and that we should keep it confidential for some time to come for various important reasons."[33] Because of the secrecy imposed on the X modulation project, no one outside the RCA organization learned about the Manila broadcast, the most impressive feat of FM technology until 1936.

Armstrong also figured into the question of whether and how to announce publicly RCAC's FM achievements. Despite the Schmeling-Stribling broadcast, Clarence Hansell realized that the company still could produce neither marketable apparatus nor even material proof that could substantiate claims of priority should someone else file a patent interference. "Circumstances," Hansell explained to Harold Beverage on 6 January 1932, "prevent our establishing credit for the development by immediately using [frequency modulation] commercially." But he feared that his team had invested and accomplished too much to chance the "great danger that some of our competitors may take most of the credit from us by prior publication." He therefore drafted an article about the Bolinas trials that he hoped to submit to the IRE's *Proceedings*, titled "Phase and Frequency Modulation Applied to Short Wave Communications."[34]

Hansell assured his staff that the company's first FM paper would recognize the efforts of all contributors to the development of frequency modulation, particularly the "very active" role of Howard Armstrong. An early draft of the article said as much. "The tests were observed closely by Armstrong, Beverage, Crosby, Hansell and Peterson," read the introduction. "The purpose of this paper is to present their conclusions together with some of the considerations and observations upon which the conclusions were based."[35] Hansell even suggested to Harold Beverage that "since Major Armstrong took such an active part . . . you will, no doubt, wish to have him pass upon the paper, or perhaps become a joint author."[36]

Unfortunately, an unknown staff attorney—probably Harry Tunick, who headed the New York office of the RCA Patent Department—quashed the article lest it spur rival firms to accelerate their own FM research and development programs.[37] Today, an examination of the FM patents that emerged during the next several years exposes those fears as groundless. In fact, RCA had sprinted too far ahead for anyone to catch up. During the three years before Hansell's proposal, only one FM-related patent per year had been issued to non-RCA inventors, while RCA had filed thirteen in the same period.

Nearly half a decade later, after Armstrong had severed his relationship with RCA, the *Proceedings of the Institute of Radio Engineers* finally published a report, in 1936, on the New York–California trials, a dry, workmanlike piece authored solely by Murray Crosby, who altogether ignored Armstrong's participation.[38] Sadly for Crosby and his employer, this article, RCA's first about FM, failed to attract the recognition it deserved. Crosby had, after all, described in detail the first nearly practical FM system, which antedated Armstrong's by two years. Moreover, probably no one understood the mathematical theory behind frequency modulation better than Crosby did. But he fell victim to the worst possible timing; the previous month's issue of the *Proceedings* contained Armstrong's celebrated paper about *his* FM system, a relatively breezy piece that diverted a great deal of attention from Armstrong's former collaborators.[39] Today, almost every historian of FM radio cites the Armstrong article, but no one mentions Crosby's.

That Howard Armstrong stole RCAC's thunder was multiply ironic. Clarence Hansell had feared in 1932 that a competing firm would beat RCA at publishing the first major article about frequency modulation. As things turned out, no such competition existed at the time, but the fear materialized anyway because of an insider—Armstrong—whose thinking about FM owed much to RCA's research. Armstrong borrowed language from Hansell's paper, for example, when his first FM paper provided the same misinterpretation of Carson's work. (Armstrong quoted the same sentences that Hansell did in support of that misreading.)[40] Also, by suppressing publication in 1932, the RCA Patent Department gave up what would have been the best insurance against the ensuing distortions of RCAC's role in the history of FM radio's development.

How differently things might have turned out—for RCA, for Armstrong, and for the way the history of FM has been told—had Hansell's manuscript seen publication before 1934, before the Armstrong-RCA relationship soured. Hansell had written the first draft of history, the sole contemporary narrative of his company's work with FM, and the only version ever to originate within RCA that has admitted to Armstrong's involvement in the Bolinas trials. The article Clarence Hansell wanted to publish would have documented that fact, and when the time came to author competing narratives of FM's development, neither Armstrong nor the company could have ignored or even denied that collaboration as they both would attempt to do.

CHAPTER FOUR

The Serendipitous Discovery of Staticless Radio, 1915–1935

> Serendipity a very expressive word. . . . You will understand it better by the derivation than by the definition. I once read a silly fairy tale called The Three Princes of Serendip: as their highnesses travelled, they were always making discoveries, by accident and sagacity, of things which they were not in quest of.
>
> *Horace Walpole, 1754*

To explain how Armstrong invented something resembling modern broadcast FM radio, I have examined how Armstrong profited both indirectly and directly from the results of thirty years of work by other men. He relied on his insider's knowledge of RCAC's research, and that firm had earlier learned much from the previous efforts of KDKA engineers and from the extensive use of Valdemar Poulsen's arc oscillator-based radiotelegraph system. In this chapter, I take a closer look at how Armstrong stumbled on what wideband FM radio actually did.

At all times, Armstrong traveled a path of invention that is best described as *serendipitous*—not in the modern dictionary meaning but in the sense that Walpole used when he coined the word in 1754. Before the issue date of the wideband FM patents, and for several years afterward, no one, not even Armstrong, completely understood the potential of what he had invented. He had based his patents on imaginative yet flawed theories of radio communication, and he therefore anticipated virtually none of the now well-known properties of modern FM radio, most notably its abilities to suppress static and reject interstation interference. To Armstrong's credit, though, he possessed the sagacious quality of intellectual adaptability, which enabled him (and others) to realize, over the course of several years, that he had invented a kind of FM far more valuable than he had intended. Moreover, when his system contradicted his expectations, he pragmatically overhauled his theoretical framework and continued to improve the technology. Certainly, this story of discovery was serendipitous in the way dictionaries define the word today—namely, as a happy accident. But the original meaning of the word fits better. In the "fairy tale" of Horace Walpole, the king of Serendip hired the best tutors for his three sons' educations. When they reached manhood, the king exiled the princes to wander the countryside, where they exercised their wits by analyzing evidence to solve a series of mysteries they happened on. Yes, luck played a part in their accomplishments, but so did cleverness. As Walpole would recognize, FM radio also resulted from a similar combination of accident *and* sagacity, not from accident alone.[1]

This chapter revises a second aspect of FM's history; namely the interaction of the evolution of FM radio technology and the competitive-cooperative relationship between Armstrong and the Radio Corporation of America. The fact that Armstrong kept much of his FM work secret from RCA adumbrated the rupture of that relationship. But shortly before or immediately after receiving his patents in December 1933, he disclosed them only to that company, which for more than a year afterward contributed considerable human and material resources for testing his system. Unfortunately, serious defects marred these tests. RCA and Armstrong designed them to investigate only the principal claim of Armstrong's patents: that FM extended the geographical range of radio waves transmitted at 30 megacycles or higher. Thus, only a small number of RCA engineers, and even fewer managers, adequately grasped what now seems like FM's obvious ability to suppress static. Furthermore, until the end of the 1930s, no one—not even Armstrong—knew about FM's capacities to reproduce high-frequency sound with superior fidelity and to resist interstation interference. Consequently, RCA engineers and managers never possessed sufficient information to make intelligent guesses about the potential of the wideband system.

The Balanced Amplifier and Frequency Modulation before 1934

At first, evaluating Armstrong's work apart from RCA during the twenties and early thirties seems especially difficult. The inventor habitually neglected to keep records of his research, confided in almost no one, and often unveiled his creations only after securing their patents. Indeed, in a rare criticism of his subject,

Armstrong's biographer, Lawrence Lessing, characterized the fact that "he was secretive and stubborn in the direction of his own affairs" as his "most serious fault." Armstrong, Lessing wrote, "would never keep a regular or orderly laboratory notebook, describing his experiments as he made them, preferring to keep everything in his head until he was ready to make a full disclosure to the world." Lessing also believed that Armstrong's secretive tendencies were "so inextricably woven into his inventive nature as to have the look of fate about them."[2] In Lessing's telling, wideband FM seems the spontaneous outcome of ineffable genius.

Yet Armstrong's patents, published articles, and correspondence confirm his confession that he arrived at wideband FM only after a long period of trial and error, during which he followed "more will-o-the-wisps than I ever thought could exist."[3] By mid-1931, when Crosby, Hansell, and Beverage came to his laboratory at Columbia University to observe his latest FM receiver, he still had not yet hit on the idea of widening the frequency swing within range of his high-fidelity system of the late 1930s, 150 kilocycles. But he was well positioned to move in that direction, chiefly on account of his nearly two-decades-long, often-quixotic quest to suppress radio noise with a circuit called the balanced amplifier.

Every experienced radio designer in 1930 knew about the balanced amplifier.[4] Recognizable in schematic drawings by its characteristic symmetry, the circuit boasted two parallel mirror-image signal paths. It was normal practice to provide each path with separate input and output connections, and to combine the two signals, at either the inputs or the outputs, for the purpose of electrically adding or subtracting the signals. A crucial advantage of the balanced amplifier was its improved linearity compared to single-path circuits, something known even before the advent of vacuum tubes during World War I. In 1915 John R. Carson used balanced amplifiers in the patent that introduced single-sideband suppressed-carrier modulation (fig. 17), and ten months later another Carson patent used a balanced amplifier in "an improved detector in which distortion is largely eliminated" (fig. 18).[5] Clarence Hansell, RCAC's FM patent workhorse, similarly employed balanced amplifiers in at least eleven patents he filed before 1931.[6] Although no one seems to have understood *why* balanced amplifiers appeared to produce "cleaner" signals than single-path amplifiers, radio designers nevertheless pragmatically made the circuit part of normal practice by the mid-1920s. As for Armstrong, balanced amplifiers entered his design vocabulary during the beginning of his career. Soon after he graduated from college in 1913, he submitted an article to the *Proceedings of the Institute of Radio Engineers*, which published the piece in 1915. In it he introduced a circuit he described as a combination of "the two most effective static eliminators known; the balanced valve

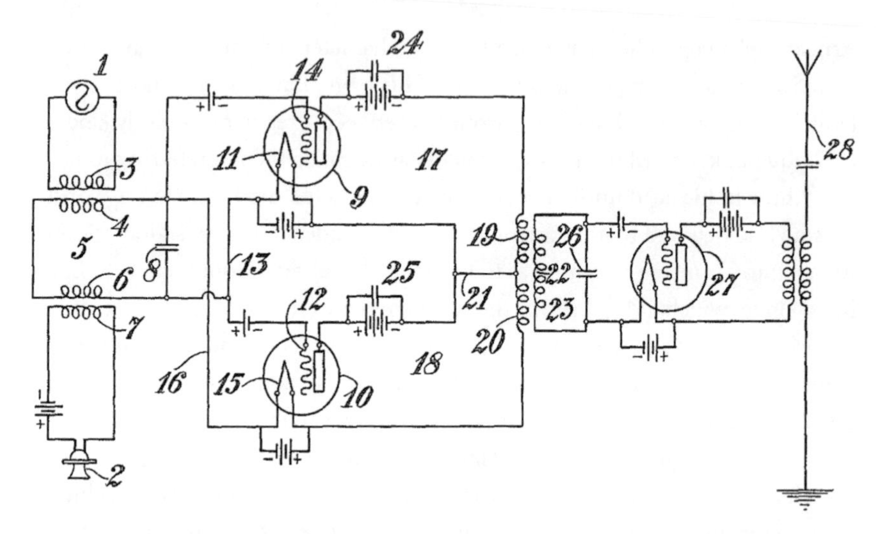

Fig. 17. Carson Single-Sideband Transmitter, with Balanced Amplifier, 1915. John R. Carson, “Method and Means for Signaling with High-Frequency Waves,” U.S. Patent No. 1,449,382, application date: 1 December 1915, issue date: 27 March 1923, assigned to AT&T.

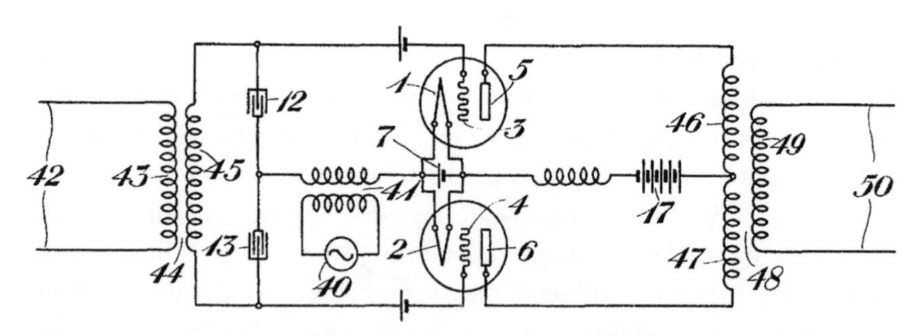

Fig. 18. Carson Balanced Amplifier, 1917. John R. Carson, “Duplex Translating-Circuits,” U.S. Patent No. 1,343,307, application date: 5 September 1916, issue date: 15 June 1920, assigned to AT&T.

and the heterodyne receiver.” As the symmetry visible in figure 19 shows, this circuit was a kind of balanced amplifier.[7]

Armstrong is sometimes mistakenly portrayed as a purely practical engineer, which has fostered the almost universally accepted misconception that theory had nothing to do with how he approached invention and design. In fact, he was a bold spinner of theories, many about balanced amplifier circuits. In 1914 his first published paper proposed a theoretical model describing how the au-

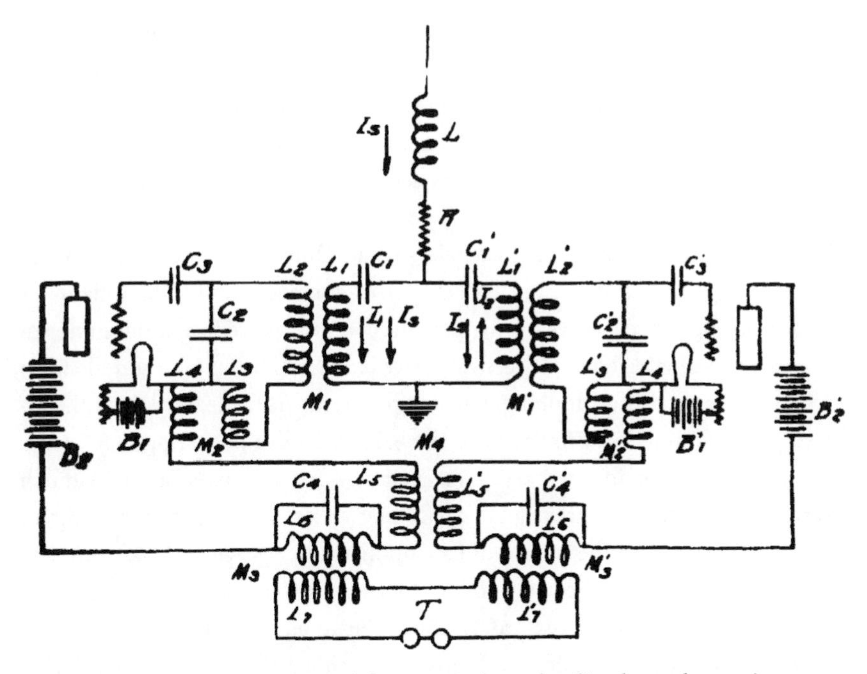

Fig. 19. Armstrong Balanced Amplifier, 1915. Balanced radiotelegraphy receiver described in Armstrong's second article, published in 1915. With "each receiver," he claimed, "it is possible to balance out the static and at the same time secure an additive response of the signals from each receiver." Edwin H. Armstrong. "Some Recent Developments in the Audion Receiver." *Proceedings of the Institute of Radio Engineers* 3 (September 1915): 215–47, reprinted in John W. Morrisey, ed., *The Legacies of Edwin Howard Armstrong* (n.p.: Radio Club of America, 1990), 67; also reprinted in *Proceedings of the IEEE* 85 (April 1997): 685–97.

dion worked, based on experiments he had carried out during his senior year at Columbia University. When the audion's inventor, Lee de Forest, ridiculed the model, Armstrong's rebuttal exposed de Forest's shaky understanding of his own invention.[8] Additional articles and patents authored by Armstrong, including his now-celebrated 1936 paper on wideband FM, also contained theories, several about the operation of balanced amplifiers. Armstrong fell short as a theorist, however, in one major respect: he never proved anything mathematically, in the style of, say John Carson or Murray Crosby, and math above the level of even high school algebra rarely appeared in his dozens of published papers. Not that Armstrong resented mathematics or mathematicians, but he tended to restrict his use of math to charts of recorded data. Perhaps his middling skills or evident incuriosity about the subject accounts for this inclination.

His overreliance on direct observation and intuition, and his underutilization of mathematics, go far in explaining his tortuous, twenty-year, and usually poorly conceived pursuit of a theory that suggested that balanced amplifiers might solve the problem of radio noise. As a college student, likely after visually inspecting paper oscillograph tapes and using his sense of hearing to compare balanced amplifiers with single-path amplifiers, he came to suspect what practitioners now share as common knowledge: that balanced amplifiers preserve fidelity better than single-path amplifiers. But no competent engineer would today claim, as did Armstrong for several years, that balanced amplifiers reduce random noises such as atmospheric static. He seems to have based this belief on a combination of hunches and wishful thinking. His 1915 paper, for instance, explained the alleged static elimination properties of the balanced valve by stating that "strays which cause serious interference [strong static-noise impulses] are of a much greater amplitude" than the signal, a fact that causes a single-path receiver to decrease its current in the output of its amplifier tube. By wiring "two complete receiving systems" in mirror fashion so that the output of one tube was subtracted from the other, he asserted, "it is possible to balance out the static and at the same time secure an additive response of the signals from each receiver." This theory rested on two precarious assumptions: that radio waves carrying static, as well as signal waves created by transmitters, always behaved (for reasons that Armstrong neglected to provide) more differently than alike in terms of the noise they caused in receivers; and that "static of large amplitude does not interact with the local frequencies [the received signal]."[9] In fact, the former assumption is sometimes true, and the static that bedevils AM radio reception today testifies to the falseness of the latter.

That the balanced amplifier captivated Armstrong's imagination is apparent also from his many patents and published articles. These reveal a line of descent from his theoretical confusion about the balanced-valve AM-detector circuit of 1915 to his low-static FM receiver of 1933 and eventually to the high-fidelity FM receiver of the late thirties. At all points Armstrong used balanced amplifiers, although he continually adjusted his explanations for how they worked, and what purpose they served. The 1915 circuit, for example, resurfaced in 1917, when he and his mentor at Columbia, Michael Pupin, invented a "Radioreceiving System Having High Selectivity" (fig. 20).[10] The two men claimed no advantage for the circuit, but in 1922 Armstrong revived his static-reduction theory for balanced amplifiers when he filed an application for a "Wave Signaling System" (fig. 21). The patent that was issued, No. 1,716,573, referred to the balanced amplifier as it was described in his 1915 audion paper (see fig. 19). Armstrong still believed in the

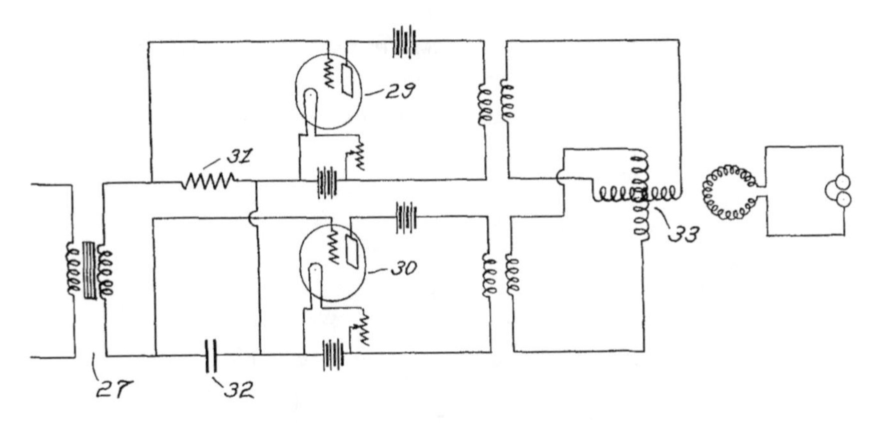

Fig. 20. Armstrong-Pupin Balanced Amplifier, 1917. Michael I. Pupin and Edwin H. Armstrong, "Radioreceiving System Having High Selectivity," U.S. Patent No. 1,416,061, application date: 18 December 1917, issue date: 16 May 1922.

static-reduction properties of balanced amplifiers, for the patent described a circuit "whereby the undesirable effects produced by atmospheric disturbances or other types of interference [i.e., static], in the course of the reception of signals, are greatly reduced." He also admitted that the 1915 circuit had "not been found to operate satisfactorily." He said this not to reject his static-reduction theory, though, but rather to explain how his new invention remedied other problems associated with the earlier circuit.[11]

The years 1927 and 1928 were the toughest during Armstrong's struggle to form a theory to explain how the balanced amplifier might suppress static. In August 1927 he filed a patent application for a low-static radiotelegraph system, and five months later he published an article about this invention in the *Proceedings of the Institute of Radio Engineers.* Again elaborating on his 1915 theory, Armstrong asserted that the energy of static noise changed significantly with respect to time but hardly at all with respect to frequency. Therefore, he concluded, more or less identical noise would inhabit two adjacent and extremely narrow channels on the spectrum at all points in time. Armstrong proposed using FSK to send and receive radiotelegraphic messages, shifting the transmitter frequency between closely spaced upper and lower channels. During transmission, one channel would contain signal + noise, and the other channel only noise—presumably noise identical to that which distorted the other channel's signal. Then he used the two sides of a balanced amplifier to detect the two channels separately. Electrically subtracting the output of one side of the amplifier from the output of the

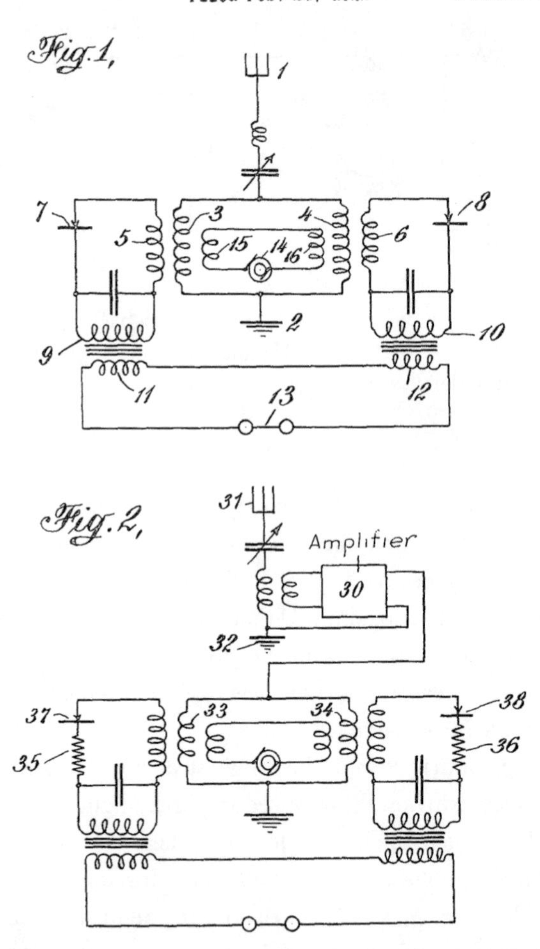

Fig. 21. Armstrong Balanced Amplifier, 1922. Edwin H. Armstrong, "Wave Signaling System," U.S. Patent No. 1,716,573, application date: 24 February 1922, issue date: 11 June 1929.

other side would cause the noise to cancel itself out (that is, signal + noise – noise = signal).[12]

This theory caught the wary eye of John Carson, who in July 1928 brilliantly cut "Armstrong's scheme" to pieces, dismissing it as merely "another arrangement which provides for high-frequency selection plus low-frequency balancing after detection." Carson proved that "no appreciable gain is to be expected from balancing arrangements" and showed that Armstrong had overstated the similarity of static noises in adjacent channels and underestimated the randomness of static in general. "In the Armstrong system," he said, "interference occurring during a spacing interval [a period of no transmission] may result in a false signal, depending on the intensity of the interference, and on uncontrollable, variable phase angles." That is, because at any point in time two channels almost always contain radically different noise, in terms of amplitude and phase, the balanced amplifier cannot cancel static simply by subtracting the noise in one channel from the noise in the other channel. "We are unavoidably forced to the conclusion," Carson declared, "that static, like the poor, will always be with us."[13] No record exists of Armstrong's immediate reaction, but he evidently accepted Carson's judgment. Save for a handful of inconsequential patents, after 1927 he ceased publishing assertions that balanced amplifiers could reduce static.

The balanced amplifier also featured prominently in Armstrong's first FM patent, a curious document that reveals him at his worst as a theorist and, ultimately, at his most serendipitous as an inventor. On one hand, "No. '447" (fig. 22) is one of only four narrowband frequency-modulation patents ever awarded by the U.S. Patent Office, all of which were filed years after John Carson debunked narrowband FM in 1922.[14] Armstrong must have known of Carson's article, by far the most widely cited of only a handful published on the subject of frequency modulation before 1935. Further, Armstrong and Carson had each authored several papers in the *Proceedings of the Institute of Radio Engineers*, published by an organization that had honored both men with the Liebman Prize for their inventions—Armstrong for regeneration, and Carson for single-sideband suppressed-carrier modulation.

At any rate, although in 1927 Armstrong did not dispute Carson's argument on mathematical grounds, he still trusted his own engineering instincts. He probably suspected for intuitive reasons that Carson's critique of narrowband FM—at least as Armstrong understood that critique—was flawed. Many years later Armstrong would implicitly acknowledge that *he*, not Carson, had been in error. But that admission came in 1935. In 1927 he still believed that narrowband FM could conserve spectrum and reduce the effects of static. "This method of modulation," he claimed in his patent, "is not subject to the usual limitations which requires [*sic*] at least 5000 cycles. The band may be made any width desired depending on the particular conditions and the distance over which it is desired to operate. This can only be determined by experiment. In general however the narrower the band the less the effect of atmospheric disturbances."[15] With good reason, Carson must have found this assertion preposterous.

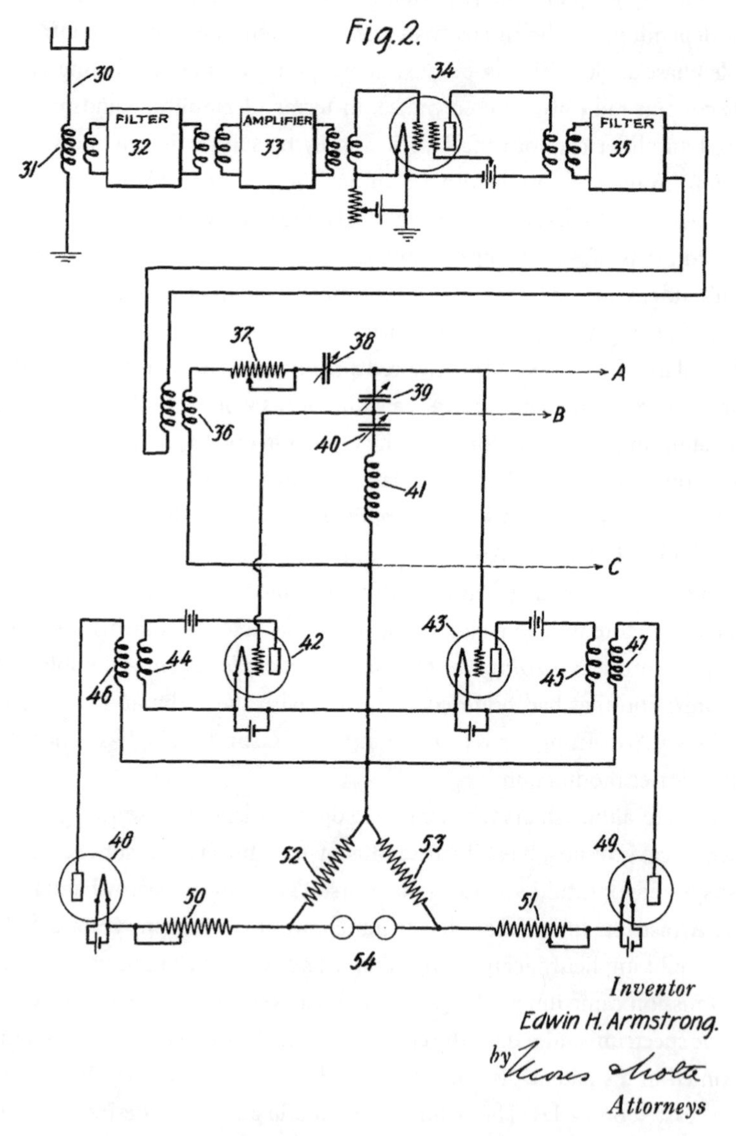

Fig. 22. Armstrong's First FM Receiver, 1927. The receiver of Armstrong's narrowband FM system. Note the balanced amplifier detector connected to the output of the filter (35). Edwin H. Armstrong, "Radio Telephone Signaling," U.S. Patent No. 1,941,447, application date: 18 May 1927, issue date: 26 December 1933.

On the other hand, No. '447 introduced a circuit Armstrong would recycle profitably a half decade later in his wideband FM system. Rather than resort to a traditional Ehret-style single-path slope detector, Armstrong instead devised a balanced-amplifier detector. One side of the amplifier was tuned to the upper frequency limit of the FM channel, the other side to the lower limit, so that the two halves worked in complementary fashion, analogous to a seesaw. Incoming radio waves at the upper radio-frequency limit caused the audio amplifier output to be driven to its negative-most amplitude. Conversely, radio waves at the lower end of the frequency swing made the audio amplifier deliver a maximally positive voltage. A radio wave with a frequency equal to the carrier's caused the audio amplifier to output a midpoint potential; usually zero volts. This design carried two advantages: first, it exhibited far greater linearity; and, second, the receiver had the potential to work over a much wider frequency swing, although Armstrong did not realize or claim this until the early 1930s.

No. '447 also marks the point at which Armstrong first began to toy with the important insight that "static, fading and like disturbances manifest themselves substantially as amplitude variations of the wave." Although he failed at the time to follow up with a crucial deduction, this statement would prove correct. Because static behaves primarily like an "amplitude variation" and not a frequency variation, an ideal frequency-modulation system, which ignores amplitude variations, resists the effects of static more effectively than an AM system does. In other words, one can minimize static distortion with two methods: either outmuscle the amplitude-distortion of the static with tremendous AM transmitter power or swamp the frequency distortion effects by maximizing the frequency swing of an FM transmitter. The former method is impractical, but in 1933 Armstrong would discover the second method and use it as the basis for low-static FM. In 1927, however, no reason existed to try extraordinarily wide swings. No practitioner had ever attempted to build a wideband system, and no theorist had considered wideband FM in the abstract. But he did believe, for at least a few months, that his *narrowband* FM receiver would resist static. Again, he cited the illusory static-reduction properties of balanced amplifiers.

Armstrong fiddled with balanced amplifiers into the early 1930s, always hoping that they would help vanquish static noise, even if they could not do so alone. His second FM patent, filed in 1930, featured an improved version of the 1927 receiver, and probably resembled the design that impressed Hansell, Crosby, and Beverage when they visited Armstrong's laboratory in late July 1931.[16] Just two months later, in September 1931, Murray Crosby filed a patent application for a similar phase-modulation receiver.[17] During the late 1930s, after Armstrong and

RCA parted ways and Armstrong sued Crosby for patent infringement, Crosby admitted that he inadvertently copied Armstrong's balanced-amplifier design during his July 1931 visit and had later failed to recall its source.[18]

Finally, on 24 January 1933, Armstrong filed applications for the two patents—Nos. 1,941,068 and 1,941,069—that laid the foundation for the high-fidelity broadcast FM radio technology that dominated the second half of the century.[19] Any radio engineer would have noticed two innovations in these documents, one largely conceptual, the other chiefly material. First, Armstrong declared the desirability of employing a "greater swing to the frequency of the transmitted wave," though he neglected to specify how much greater. His point was that widening the swing had positive benefits and was not merely a necessary expedient. Second, Armstrong permanently scrapped simpler modulation and demodulation circuits that had been recognized as normal design practice for FM technology for decades. Since 1902, all FM radiotelephony transmitters had used a variable reactance modulator, at first completely mechanical versions like Cornelius Ehret's, in which sound waves physically altered the electrical properties of reactive components such as condensers or inductors. Later, KDKA engineers introduced more sophisticated *LC*, crystal, and electronic reactance circuits that mimicked Ehret's modulator. Now, going even further, Armstrong used a balanced amplifier as a modulator. He also abandoned the reactance demodulator, introducing a circuit built around a balanced-amplifier circuit with the usual twin-signal paths, essentially a reworking of his narrowband system of 1927. But the new receiver boasted a superior linearity over far wider frequency deviations, and he used it as a "means for selecting these large swings of frequency."[20]

What Armstrong Thought He Had Invented

With a new transmitter design and a modified six-year-old receiver in hand, Armstrong had crafted a system that resembled in many ways modern broadcast FM radio. Yet, paradoxically, his patents described no such thing. Armstrong failed even to hint at two characteristics that later made FM the first high-fidelity mass medium—namely, a wider audio bandwidth that reproduced sounds far more realistically, and the ability to remove almost all static noise. Nor did he know yet that the radiation pattern of an FM transmitter was easier to regulate than that of an AM station and that frequency modulation all but erased the effects of interstation interference.[21] He did imply that his FM could convey wider bandwidths, but not in connection with audio reproduction; rather, Armstrong declared that this ability would prove useful for sending television or facsimile images.

No. 1,941,068, the more narrowly focused of the two core patents for wideband FM, straightforwardly defined the transmitter: a dual-input balanced-amplifier circuit that mixed two out-of-phase waves. Injected into one input of the amplifier was an unmodulated radio-frequency carrier; into the other input, an amplitude-modulated wave of the same frequency as the carrier but ninety degrees out of phase with the carrier. Adding the two signals electrically resulted in a composite signal at the output: an amplitude-modulated radio-frequency wave whose phase shifted in proportion to the instantaneous amplitude of the audio signal. (This technically amounted to phase modulation, but conversion to frequency modulation is a simple matter.) "Limiter" stages that followed removed the amplitude-modulated components of the wave, and subsequent "doubler" and "tripler" stages multiplied the signal frequency to the range of the transmitter frequency. These multiplier stages widened the frequency swing as well.[22]

The patent also explained that the balanced modulator worked "by aperiodic means, (that is, without the use of resonant circuits and therefore without the creation of transient oscillations therein)." By this, Armstrong meant that he had jettisoned the old reactance modulator, whose response curve conformed to that of a tuned *LC*, "periodic" circuit. Armstrong observed that *LC* circuits used this way tended, unfortunately, to "ring"—that is, to emit short-lived spontaneous, "transient" oscillations. His transmitter also used resonant circuits but not in parts of the apparatus where transients would most likely crop up.[23]

Patent No. '068 allows for little theoretical interpretation; it described only the behavior of an invention. By contrast, No. 1,641,069 staked out broad theoretical claims about radio in general and about the advantages of frequency modulation specifically. It is also laden with assertions—many of them ambiguous, unsupported, or misleading—that reveal what Armstrong *believed* he had invented. Most important, he declared that the primary objective of his invention was to "[increase] the distance of transmission which may be covered in radio signaling with very short waves."[24] That is, his system would extend the maximum geographic range of radio waves that operated at frequencies far above the standard AM broadcast band. This assertion, while largely valid, played a central role in Armstrong's failure to sell FM to RCA.

A later-discredited assumption about noise in the upper radio frequencies lay at the foundation of Armstrong's claim about FM's greater range—and his failure to anticipate FM's static reduction properties. Armstrong believed that static noise, still the bane of AM radio today, did not affect reception in certain higher-frequency ranges. "It is well known," he declared in the opening lines of '069, "that waves of the order of ten meters or lower [30 megacycles or greater] are limited in the distance of transmission by tube noise alone as *the amount of static in that part of the spectrum is negligible*."[25] This statement reflected a commonly held misconception among radio engineers during the early 1930s, when much of the spectrum above the AM broadcast band remained barely explored and poorly understood—namely, that the "ultra-high" frequencies were almost free of static. Additional experience would teach that static inhabits the ultras at comparatively low energy levels. Further, the relatively steady intensity of upper-frequency static noise (as opposed to the characteristically dynamic impulsive static of the AM broadcast band) caused engineers to suspect wrongly that static hardly existed in that region of the spectrum. Although high-frequency static does differ from AM-band static, the dissimilarities are as much qualitative as quantitative.

Armstrong coupled his belief in an almost-static-free region of the spectrum with an even more mistaken conviction that *frequency modulation can effect no reduction in static noise*, a recent version of a theory he had been continually constructing, tearing down, and rebuilding for nearly ten years. Formerly, of course, Armstrong had believed FM *could* reduce static, and his 1927 narrowband FM patent declared that he had made "an invention for eliminating the effects of fading and static."[26] But John Carson had exposed the futility of narrowband FM five years earlier. In 1930 Armstrong applied for another balanced-amplifier FM invention but stated no opinion on the question of whether FM reduces static. Now, he was pivoting 180 degrees from his original position. No. '069 wrongly asserted that static degrades the quality of FM transmissions in proportion to the channel width, just as it does with AM. "Band widths have always been kept down to as low a value as possible," he explained, "because the amount of static which is received is proportional to the width of the band."[27]

To justify this claim, he transplanted into FM theory the well-known relationship between static and channel width as it applied to AM radio: "It is the practice in designing amplitude modulated receivers," he stated, "to design the width of the selective system to be equal to twice the frequency of the modulation to be received." With FM also, he continued, "while there is no [standard] practice, the experimentation has proceeded along the same lines, the width in this case being somewhat greater than in the amplitude-modulated case in order to allow for the deviation in frequency." Armstrong went on to explain that because the bandwidth of an FM channel depends on the frequency swing, to widen the channel is to invite proportionally more static, which will degrade the signal: "These band widths have always been kept down to as low a value as possible because the amount of static which is received is proportional to the width of the band and hence after providing for the signal there is no advantage in going further." By asserting that "this applies both to amplitude and frequency modulated waves," he implied that static noise degraded wider FM channels and AM channels in the same way.[28] Ironically, the very invention whose patent contained this language later proved that Armstrong simply had it backward. Practitioners now know that broadening the channel has the opposite effect on FM compared with AM; a wider channel *reduces* static noise on the former and *increases* static noise on the latter.

If Armstrong believed in 1933 that FM per se could not reduce static noise and that widening the channel width allowed more static to distort the audio signal, what advantage did he hope to gain from a wideband system? The implicit answer to this question rested on still another misbegotten theory. Because, he asserted, negligible static exists at or above 30 megacycles, "it is well known that waves [there] . . . are limited in the distance of transmission *by tube noise alone*."[29]

Here, one must understand clearly what Armstrong meant by "tube noise"—or its synonym, "tube hiss." Tube noise was not the same as "static noise." The two were—and still are—understood to be different in terms of their origins. Sources outside the receiver, either natural ones such as lightning or man-made ones like electric motors, cause static. By contrast, the radio receiver's vacuum tubes produce the molecular- or quantum-level white noise called tube noise. As Armstrong put it, the nature of tube noise, "which is due mainly to the irregularities of the electron emission from the filaments of the vacuum tubes, is that of a spectrum, containing all frequencies."[30] This emission makes a hiss audible to human ears.[31]

Tube noise, to be sure, was a real and worsening problem in 1933. During Armstrong's youth, radios had no more than one stage of vacuum-tube amplification, sufficient to drive a low-power headset, but not enough to create especially objectionable hissing. But newer radio receivers amplified weaker signals and had to drive loudspeakers, which required more wattage than a single-stage audio amplifier could provide. Therefore, after 1920 designers began employing two or more stages of "cascaded" amplification, which fed forward the energy of an amplifier tube stage to the input of a succeeding stage. Multiplying the individual gain factors of each stage obtained the overall amplifier gain. For example, if the three stages of an amplifier boosted the input signal by gains factors of 20, 50, and 100, respectively, the overall gain was $20 \times 50 \times 100 = 100{,}000$. This design practice made for more sensitive radios, but each tube in a cascaded amplifier unfortunately amplified the accumulated noise of the preceding stages. By the third or fourth amplifier, tube noise, as Armstrong said, "manifests itself in the [speaker] by a high pitched hiss, the frequencies composing which run from some low value to above audibility."[32]

One document in Armstrong's papers demonstrates that he had tube hiss and not static in mind when he began investigating wider frequency swings for FM. On 21 July 1932 he sketched a circuit diagram of an FM receiver that closely resembled his balanced amplifier design of 1927 (see fig. 23 for the sketch and fig. 22 for a patent drawing of the 1927 receiver). At the top of the page he scrawled: "Demodulation of Tube Noise by Frequency Modulation at 7.5 meters [40 megacycles]." In the bottom right quadrant, he summarized the results of an experiment:

Frequency swing
 50 to 60 K.C.
 Comparison with amplitude
modulation showed very
many times improvement
of hiss ratio.
 Demonstrated to C. R. Runyon Jr.
July 20, 1932 at Hartley
Research Laboratory.
 (signed) E. H. Armstrong
 July 21, 1932.
 (signed) C. R. Runyon Jr.
 July 21, 1932[33]

This text reveals several aspects of Armstrong's work with FM in mid-1932. First, although a "50 to 60 K.C." swing spanned a narrower slice of the spectrum than the 150-kilocycle high-fidelity broadcast FM systems of today, Armstrong was already working with frequency deviations more than twice those RCAC engineers were using. That only four weeks later Armstrong penned a draft of one of his first wideband FM inventions, No. '069, indicates that sometime during July or August 1932, he adopted for his system an even wider frequency swing.[34]

The sketch also sheds light on Armstrong's personal loyalties, foreshadowing his increasingly profound suspicions of large organizations in general and of RCA in particular. Who else knew about Armstrong's FM development? Of course, he had to admit into his circle of confidants his patent lawyers and an assistant or two. Those men needed to know about his research in order to do their own work, and professional ethics as well bound the lawyers to secrecy. But only

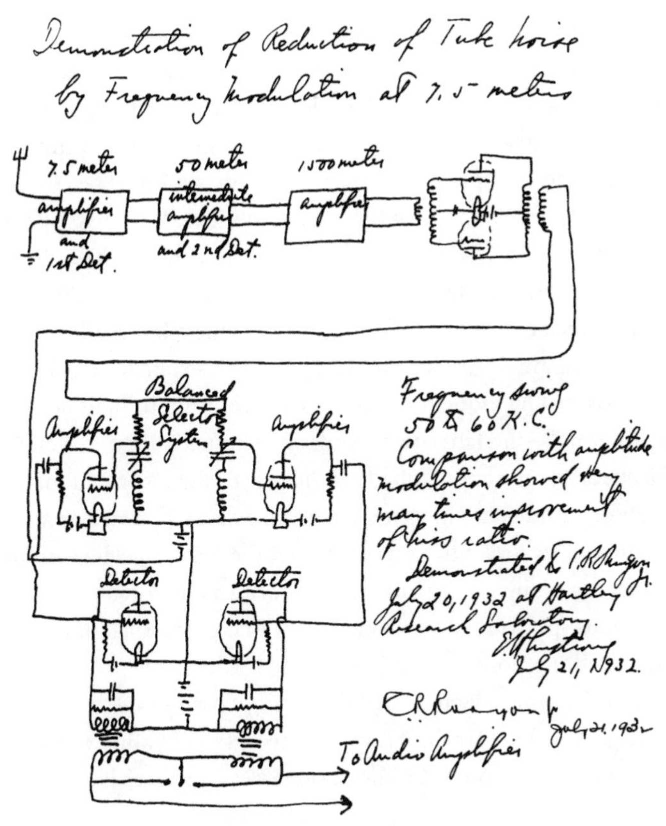

Fig. 23. Armstrong's Sketch of Wideband FM Receiver, July 1932. Edwin H. Armstrong. Memorandum. "Demonstration of Reduction of Tube Noise by Frequency Modulation at 7.5 meters," 21 July 1932, box 159, AP.

friendship can account for the signature of Carman Runyon Jr. as a witness. He and Armstrong had grown up together in the same Yonkers neighborhood, and Runyon had cofounded the Radio Club of America. In fact, Runyon had probably recruited Armstrong into that organization, where both men remained active members. Though no professional practitioner—he managed a coal delivery company for a living—only Runyon enjoyed Armstrong's implicit trust, and together the two men would stage several of the 1930's most important demonstrations of frequency modulation. By contrast, until late 1933 Armstrong confided in no RCA employee about wideband FM, not even his friends David Sarnoff and Harold Beverage, or any of the other RCAC engineers who had freely shared with him information about *their* work.

Armstrong's sketch, as well as his subsequent draft of U.S. Patent No. '069, also shows that he had both John Carson and balanced amplifiers in mind. But at this point he no longer considered the latter as a means to fight static, for Carson had taught him the uselessness of that idea. Rather, Armstrong's perception of a nonexistent clear distinction between tube hiss and static led him to believe that he had spotted a loophole in Carson's analysis, and thus he concocted still another flawed theory concerning balanced amplifiers: tube hiss, he asserted, is predictable, and static is not. He probably based this conviction on the fallacious logic that, because the sibilance of tube hiss can *seem* less random to the ear than does the crackle of static noise that plagued the AM broadcast band, then tube hiss *is* less random. Armstrong never said this explicitly, but his choice of language shows that he thought of tube hiss as "continuous" and not "irregular" or "discontinuous"—the latter two words he habitually used to describe static noise. "Electrically," he stated, tube hiss, "is practically a continuous spectrum. In this it differs from static in that static is an extremely irregular spectrum in which, because of its discontinuous character, the peaks may be commensurate with or greater than the signal before serious disturbance occurs."[35] By using the term continuous spectrum, Armstrong meant that over a span of time tube hiss generated something close to what is now called "white noise": radio waves of all frequencies at the same average amplitude, as opposed to the bursts, pops, and crashes that all too often blank out AM radio reception. In other words, tube hiss, when compared to static on AM radio, more or less *sounds* as if it lacks randomness.

This argument, which implied that a receiver could distinguish tube hiss from static, explains Armstrong's return to the balanced amplifier. He understood—correctly—that if a balanced amplifier adds two signals, one the negative of the other but otherwise identical, then the amplifier will cancel out both signals. A 100-kilocycle sine wave, for example, subtracted from an identical in-phase 100-kilocycle sine wave yields a null output. Therefore, his thinking went, subtracting the apparently "continuous" tube noise from itself will remove the noise. He had tried the same theory in 1927 to reduce static.

But Armstrong failed to comprehend that, for a balanced amplifier to work this way, the signals in both paths of the amplifier must have identical phases and amplitudes *at all points in time.* Alas, although tube hiss over the long run exhibits a more or less constant *average* amplitude over a wide spectrum, at any arbitrary *instantaneous* point in time its amplitude and phase are at least as unpredictable as garden-variety impulsive static. Indeed, tube hiss is analogous to an undisturbed blanket of snow that appears uniformly flat, but which actually comprises countless unique crystalline structures. Our sense of hearing similarly deceives us by "averaging" the sound of tube hiss so that it sounds "the same." But a modern spectrum analyzer reveals that tube hiss resembles "white" noise, the most random kind possible, and therefore the most resistant to noise-cancellation schemes.

Armstrong's ill-conceived theories sent him in a wrong direction, but ironically the same path led to the serendipitous development of modern FM radio. Because the ability to reduce tube hiss in Armstrong's system came, according to him, not from frequency modulation but rather from the balanced amplifier's properties, he probably considered using a balanced amplifier circuit to do the same thing with AM receivers as well. But he knew of two advantages FM held over AM. One was FM's demonstrated ability to resist fading. The other was his private discovery of "a very great improvement in transmission" due to wide frequency swings, which he mischaracterized as a reduction in tube noise alone.[36]

More than seventy years of hindsight reveals that Armstrong was on to something, but for the wrong reasons. However flawed his original theories behind wideband FM, he had, for the first time, hit on a reason beyond expediency for widening the frequency swing. Simply put, widen the swing, reduce the level of tube noise. And wider swings were possible only in the recently discovered expanses of the spectrum's short wave regions. But, again, he had not yet made the connection between frequency swing and static noise suppression. (That epiphany would strike him a few weeks after field tests of wideband FM began in early 1934.) Moreover, modern FM, which closely resembles Armstrong's system, *does* sound quieter—that is, less noisy—thanks to a greater frequency swing. But the inventor of wideband FM misunderstood what caused this "very great improvement in transmission" (i.e., a greater geographic range) and the reduction of noise. He believed that FM with balanced amplifier receivers reduced *only* tube hiss. And he initially interpreted the lower levels of audible and quantifiable hiss on his prototype FM receiver as confirmation of this hypothesis. Further, Armstrong's other presumption, that static virtually did not exist in the upper frequencies, was wrong as well. In fact, static noise did exist above 30 megacycles, but it resembled white noise—and tube hiss—more than it did the static crashes that polluted the AM band. A good deal of what Armstrong took for *tube*-noise reduction was probably *static*-noise reduction instead, and a wide frequency swing—not the balanced amplifiers—actually accounted for virtually all the reduction of both kinds of noise.

RCA and Armstrong's Discovery of Staticless Radio, 1934

Sometime around the issue date of his patents on 26 December 1933, Armstrong began disclosing his system to a handful of RCA employees by transmitting FM signals across his Columbia University laboratory. Precisely which outsiders witnessed the first of these presentations is uncertain. In 1939 he remembered that he selected the president of RCA for that privilege. "In December 1933," stated Armstrong, "I gave Mr. Sarnoff a demonstration of my system at Columbia University, following it up during the next two months with demonstrations to some two dozen or more of the leading engineers of the Radio Corporation and its subsidiaries."[37] No one recorded Sarnoff's impression of this event, and in any case Armstrong may have revealed his invention even earlier to Harold Beverage, who recalled (also in 1939) that "shortly before his patents issued, [Armstrong] disclosed his wide band frequency modulation [to me]."[38]

The reactions of the RCA engineers who visited Armstrong's lab in January 1934 ranged from guarded optimism to pure enthusiasm. Significantly, however, virtually no observer, including Armstrong, described hearing anything close to modern FM radio. On 3 January 1934 Murray Crosby wrote in his notebook that "Bev [Harold Beverage], Pete [Harold Peterson], Hansell and I went to New York to a demonstration by E. H. Armstrong of the system of his U.S. Patent 069." Crosby, the world's preeminent frequency-modulation theorist at the time, prudently neither endorsed nor challenged Armstrong's contention that FM decreased tube noise levels. But he did note that "by using this high [frequency] deviation and a broad band receiver," Armstrong claimed "that more noise is eliminated in the output."[39] This was a fair, even generous assessment at the time, given that no one besides Armstrong had field-tested the system. Six months later Harold Beverage held a rosier view of the same demonstration. He recalled in June that "we agreed that there was a large improvement in the signal to noise ratio in the laboratory demonstration, on the order of 15 or 20 fold."[40] Because RCA had not yet constructed equipment for quantitatively assessing the Armstrong system, though, Beverage either estimated these figures or projected them back from observations made later.

These demonstrations soon led to RCA's most direct material contribution to the development of wideband FM radio. For the better part of two years, the company would lend the inventor one of the best-equipped broadcasting test laboratories in the world. NBC had recently installed four experimental "ultra high-frequency" transmitters in the crowded upper two stories of the "newly-completed but sparsely-occupied" Empire State Building.[41] RCA constructed these stations, which operated on frequencies around 41 megacycles, primarily for the development of broadcast television, and thus named the site the "Empire State Building Television Laboratory." But the network's engineers tested other communications technologies there as well. RCA's corporate report for 1933 declared, for example, that by exploiting the "ultra high frequencies" the company "proposes to introduce the first domestic facsimile radio communication service between New York and Philadelphia." Another project was the development of "multiplex transmission," the "simultaneous sending of three different radiograms on one wave-length."[42]

Harold Beverage played a central role in obtaining a place for wideband FM in the television laboratory. For years he had been attempting without success to obtain lab space and funding to field-test RCA's earlier versions of FM. In April 1932 he wrote Charles Young, a senior engineer in RCA's recently created manufacturing arm, RCA Victor, and Charles Horn, the chief engineer of NBC (and KDKA's chief engineer until 1929), about "the possibilities of [developing] high quality broadcasting on the ultra short waves," which Beverage explained, "[applied] frequency or phase modulation to the ultra short wave transmitter." Beverage implied that FM suppressed static with balanced amplifiers. "Frequency modulation," Beverage said to Young, "can be used to very great advantage over circuits where selective fading is not a factor. . . . It has been found that frequency modulation with a properly designed receiver, will balance out a great deal of man-made interference such as automobile ignition, power noises, etc." "Major Armstrong has been working quite closely with us on this problem," he added. Why Beverage failed to secure the Television Lab for FM radio tests in 1932 is uncertain, but his statement that "Mr. Horn . . . is apparently favorable to the project if it can be financed" suggests economic, not technical reasons.[43]

Two years later, Beverage made essentially the same request, but he secured a favorable answer. On 12 January 1934 Beverage took Armstrong to the Television Laboratory "to discuss some experimental work [with FM] which Major Armstrong expects to do here in the near future."[44] Beverage sold Armstrong's new system to Horn as a noise reduction technology that demanded field-testing, "as it is always easy to be fooled by laboratory demonstrations."[45] Horn agreed, and from March until June, Armstrong and two assistants installed the first wideband FM broadcast transmitter in the Empire State Building.

It was Howard Armstrong's bad luck at this most hopeful stage of wideband FM's development to be forced to bear the greatest and unfairest personal setback of his life. The story begins in 1912, when he invented the "feed-back" or regenerative circuit. This device, his first invention, quickly became one of the most commercially important circuits in the field of electronic engineering. Even today virtually all electronic devices contain a number of feedback circuits. But in 1912, Armstrong was still a college student, and he lacked the $150 necessary to file a patent application, and so in January 1913, he resorted to paying twenty-five cents to have a sketch of his circuit notarized. Not until the following October did he scrape up the cash to file a proper application.

That nine-month delay set a tragedy in motion. Soon, Armstrong faced three patent interference claims. The courts threw out two, but that of Lee de Forest survived because de Forest documented his challenge with a laboratory notebook entry for a similar audion-based regenerative circuit, dated five months earlier than Armstrong's notarized sketch. The stakes escalated when large corporations took the side of each man. At first, AT&T, which had bought the rights to all audion-related inventions, stood in de Forest's corner, and when AT&T transferred its radio patents to RCA, the latter company, following the logic of ownership, also sided with de Forest. Armstrong found a backer in Westinghouse, which retained the rights to *his* version of regeneration.[46] But after the RCA "Radio Group" bought the rights to Westinghouse's radio patents, litigation among the corporations involved ceased.[47]

In 1931 Armstrong reopened the case, this time against de Forest personally, in hopes of winning punitive damages. By then, virtually all engineering authorities had judged the dispute as an open-and-shut case in favor of Armstrong. They pointed out that de Forest's notebook entry related only to *audio*-frequency waves in wired circuits, which aside from being electrical, had no connection to radio waves. Thus, the Institute of Radio Engineers, by awarding Armstrong its first Medal of Honor for regeneration in 1918, effectively rebuked de Forest, despite the fact that both men enjoyed considerable prestige in the organization. As the case dragged into the late 1920s, Lessing writes, "all Armstrong's amateur and professional friends, who had lived through the facts of the case, rallied to the cause, which became a leading and sulfurous topic at meetings of the Radio Club of America and the Institute of Radio Engineers."[48]

The months passed as the case worked its way up a ladder of appellate courts. Finally, on 21 May 1934, just as Armstrong was installing his FM modulator in the Empire State Building, Supreme Court Justice Benjamin Cardozo effectively ruled in favor of de Forest by affirming the decision of lower-court judges. Inexplicably, Cardozo failed to grasp that they had wrongly conflated de Forest's audio-frequency waves and Armstrong's radio-frequency waves. Exactly one week after Cardozo's decision, and two months after Armstrong began working in the NBC Television Laboratory, the humiliated inventor stepped before "nearly a thousand engineers" who had assembled for the annual meeting of the Institute of Radio Engineers. He intended to return his Medal of Honor, which he had won for regeneration, by reading a prepared statement:

> It is a long time since I have attended a gathering of the scientific and engineering world—a world in which I am at home—one in which men deal with realities and where truth is, in fact, the goal. For the past ten years I have been an exile from this world and an explorer in another—a world where men substitute words for realities and then talk about the words. Truth in that world seems merely to be the avowed object. Now I undertook to reconcile the objects of these two worlds and for a time I believed that that could be accomplished. Perhaps I still believe it—or perhaps it is all a dream.[49]

Armstrong never completed this speech. IRE president Charles Jansky Jr. (who in 1938 would establish Washington, D.C.'s first FM broadcast station) cut Armstrong off with an announcement that the institute's board of directors "hereby strongly reaffirms the original award, and similarly reaffirms the sense of what it believes to have been the original citation." Accordingly, the board of directors, "half of whose members," says Lessing, "were prominently employed by A.T. & T. and R.C.A. or their affiliated companies"—firms that had taken de Forest's side in court—refused Armstrong's offer to return his medal in dishonor.[50]

Armstrong should have taken comfort from his colleagues' expression of moral support and let the matter go. Many successful independent inventors in the early twentieth century coped with the exasperating distractions of patent litigation by leaving legal matters to their attorneys. But few patent fights in American history have drained the disputants more than the regeneration case did, and Cardozo's ruling permanently scarred Armstrong psychologically. After 1934 he continually narrowed his circle of those he trusted and for the remainder of his life cultivated a cynical worldview that profoundly influenced how he developed and promoted wideband FM radio. At first, he directed his contempt chiefly toward lawyers, who lacked, in his opinion, both the special knowledge and moral integrity necessary to evaluate technological issues. Later, as he perceived unwarranted resistance to wideband FM, he folded into his list of enemies the Federal Communications Commission, Congress, RCA, and his friend David Sarnoff. Eventually, to disagree with Armstrong about any point of radio technology was to invite having him, his allies, or his biographer impugn one's competence and character.

Even while reeling from his defeat, Armstrong persevered at the hard work of installing his system in the Empire State Building lab. On 26 March, NBC engineer Robert Shelby reported, Armstrong began calibrating "all [radio-frequency] stages and the antenna system."[51] Armstrong and the handful of RCA and NBC engineers assigned to help him had to integrate his home-built FM modulator with a 2,000-watt General Electric transmitter originally designed for television broadcasts. On one occasion, an acute electrical mismatch through a 275-foot-long coaxial cable connecting the transmitter and antenna almost wrecked the job. The cable ran along the exterior of the skyscraper, so replacing the line was out of the question, but Philip Carter, one of Armstrong's assistants, "completely solved" the problem mathematically, as Armstrong described, "in a very beautiful manner."[52] The task of mounting a bank of resistors that were designed to absorb power from the transmitter proved almost as challenging.[53] Diverting to the bank a portion of the wattage that otherwise would go to the antenna enabled Armstrong to reduce the radiated power to as little as 20 watts.

During these months, Armstrong experienced one of the most serendipitous moments in the history of radio technology. He began to realize that wideband FM radio suppressed static, a finding that squarely contradicted the language of the same system's patents. One NBC engineer later described Armstrong "as sometimes unduly secretive about his objective or the nature of the problem he was trying to solve," so perhaps predictably Armstrong never explicitly acknowledged to others his change of mind.[54] But he clearly made it, perhaps all of a sudden, but more likely after dozens of hours of tests in which he transmitted speech from the television lab to his receiver in Philosophy Hall a few miles away, and probably to Carman Runyon's home in Yonkers as well.[55]

Armstrong spread word of his discovery to RCA and NBC employees and also likely allowed some people to hear for themselves. Curiously, a lawyer first recorded in writing the new finding. Harry Tunick, who headed the RCA Patent Department's New York office, declared in May that "the essence" of the Armstrong system "resides in the reduction of noise, mainly atmospherics [i.e., static]." Armstrong, he explained,

> has made a decided advance in the art by showing that with ordinary frequency modulation systems in which the carrier is swung at the modulating frequencies over a small range corresponding, say, to the present day permitted 10 kilocycle range of frequencies, atmospherics and noise will come through. However, [Armstrong] is the first to teach the thought that by exceeding this range so that many more sidebands are required in the detector to reproduce the signal, noise components spread themselves out over this frequency spectrum in such a way as to become self-canceling.[56]

These words almost exactly conformed to Armstrong's peculiar new theory of FM. He now believed that the noise components "spread themselves out" over the channel and somehow "self-canceled." Tunick also made distortions of record that Armstrong—and later Lessing—would ultimately incorporate into what became the canonical history of FM. In fact, contrary to Tunick's (and Armstrong's) assertion, no one had attempted to design an "ordinary frequency modulation system" with a 10-kilocycle frequency swing in more than a decade. Further, Armstrong's patents decidedly did *not* "teach" that wider frequency swings accounted for FM's static noise-reduction capability. On the contrary, the patents unequivocally implied that wider swings *increased* static. But Tunick correctly said that the Armstrong system itself reduced static, a belief that he could have learned only from direct observation or discussing FM with the small number of men who were privy to Armstrong's work.

The Westhampton Beach Demonstration of 12 June 1934

As the summer of 1934 approached, Armstrong prepared to prove to a wider audience his system's fitness as a low-static, long-range radio medium. In early June he completed the transmitter's installation, and on the afternoon of the twelfth, more than half a dozen RCA engineers and managers congregated at the Westhampton Beach, Long Island, summer home of George E. Burghard to witness the first field tests of wideband FM. Burghard, a veteran ham operator, owned "a modern amateur station with all facilities," located eighty-five miles from the television lab, and eight or nine hundred feet below the line of sight of the Empire State Building.[57] He also had cofounded the Radio Club of America and, like Armstrong, had served a term as the club's president.[58]

Not everything went smoothly, but when problems cropped up, Armstrong improvised. A few days earlier, he had hauled the components of his prototype FM receiver to Burghard's home and set up the apparatus on half a dozen tables. He and Philip Carter also strung a large *V*-shaped antenna on a frame of sixty-foot-long wooden poles, an arrangement intended to maximize the signal strength. Clarence Hansell, one of the attendees, pronounced this antenna an "excellent" design, but a technical glitch almost wrecked the show. Absentmindedly, Armstrong had designed a horizontally polarized receiver antenna; that is, it was physically oriented parallel to the surface of the earth. The transmitter antenna mounted on top of the Empire State Building radiated incongruous vertically polarized waves, making for the worst possible alignment for reception. Because neither antenna could be rotated, Armstrong disconnected the large-frame antenna and hung in its place "a vertical piece of bell wire about 10 to 12 feet long, inside [Burghard's] house."[59] Although much shorter and therefore less sensitive than the original *V*-design, the makeshift aerial seemed to do the job.

Historians of FM radio have credited the Westhampton Beach demonstration with providing incontrovertible evidence of the superiority of the Armstrong system. Armstrong's biographer, Lawrence Lessing, declares that the Westhampton test, which he did not witness, proved Armstrong's FM "to be something even beyond his own expectations." Lessing also quotes an entry in George Burghard's ham radio station logbook that prophesied "an era as new and distinct in the radio art as that of regeneration [i.e., feedback]."[60] But the tests actually produced ambiguous results that concealed FM's ability to suppress static, for reasons that hindsight makes clear. Armstrong's patents maintained that an extended geographic range constituted the primary advantage of his system, so not only RCAC engineers but Armstrong as well understandably desired to determine the system's maximum range. Since then, however, it has been determined that normal reception requires a signal strength at least twice the strength of the ambient noise that exists in the channel. Listeners at the periphery of a normal-listening radiation zone typically hear surges and fading in signal strength—much like what many of those who attended the Westhampton tests reported. Thus, because no one recorded replicable quantitative data, the Westhampton Beach tests amounted to an audible inkblot test.

Not surprisingly, Armstrong proffered the most positive interpretation, one that eventually found its way into the canonical history. Two years later, in his seminal 1936 paper on wideband FM, he stated that the Westhampton tests "surpassed all expectations. Reception was perfect on any of the antennas employed, a ten-foot wire furnishing sufficient pickup to eliminate all background noises." He added that "the margin of superiority of the frequency modulation system over amplitude modulation . . . was so great that it was at once obvious that comparisons of [AM and FM] were principally of academic interest." Armstrong also described how, when the Empire State Building transmitter was reduced in power from 2,000 to 20 watts, the observers heard a "signal comparable to that received from the regular [AM] New York broadcast stations (except WEAF, a fifty-kilowatt station located approximately forty miles away)." "Under all conditions," he continued, "the service was superior to that provided by the existing fifty-kilowatt stations, this including station WEAF." "During thunderstorms," he explained, "unless lightning was striking within a few miles of Westhampton, no disturbance at all would appear on the system, while all programs on the regular broadcast system would be in a hopeless condition."[61]

Among RCA engineers at the time, however, only Harold Beverage approached this level of optimism, in a report he sent to C. H. Taylor, RCA's chief engineer. Beverage was a dubious source of reliable information, though, because a "tonsil operation" had prevented him from actually going to Westhampton. But Beverage already knew about FM's ability to reduce static because he had heard several tests of FM in Manhattan, at ranges short enough for the Armstrong system to operate under what would resemble normal conditions today. He had also discussed FM with Howard Armstrong, who "gave me copies [of his patents] and [had] demonstrated his laboratory setup to Messrs. Hansell, Peterson, Crosby, and myself" the previous January. Beverage completely accepted Armstrong's new theory that a wider swing accounted for FM's presumed suppression of static noise. There are, he told Taylor,

> inherently considerable noise reduction possibilities in the use of frequency modulation in ordinary ways on ultra short waves, as we have found by our own tests. Armstrong carries this reduction further by using a much greater swing than is ordinarily used. We have not determined accurately, just how much additional noise reduction is produced by the greater swing, but Armstrong believes it is proportional to the swing, and is probably correct in his analysis. . . . Major Armstrong said that the [Westhampton Beach] demonstration [of 12 June] was very successful and that while WEAF reception was ruined by static from local thunderstorms, the same program via the Empire State was excellent, practically free of any kind of noise.[62]

Beverage did admit that "I have not had an opportunity to talk with any of our engineers to determine their reaction. Neither do I know," he added, reflecting the absence of quantitative test results, "how much of the noise reduction may have been due to the characteristics of the ultra short wave and how much due to Major Armstrong's invention." He also acknowledged the system's "greatest drawback"—namely, "the wide frequency band required," but Beverage pointed out that "wide bands are inherently available on the high frequency end of the ultra short wave band." He concluded his letter by presciently recommending the development of FM as a commercial medium. He predicted that the Armstrong system "should be of great value in certain fields, particularly the sound entertainment field." "At least," he offered, "it is worth our while to make some investigation of the possibilities."[63]

Virtually no one else shared Beverage's and Armstrong's enthusiasm, however. A few days later, Clarence Hansell reacted tepidly, partly because his theoretical preconceptions about frequency modulation prevented him from swallowing Armstrong's new ideas. Five days after Beverage posted his memorandum to Taylor, Hansell (who received a copy of the letter) stripped the veneer from Beverage's analysis, first by reminding Beverage that he (Beverage) had not actually attended the Westhampton Beach test. "You indicate that you have not been fully informed," he stated. During the evening of the twelfth, Hansell continued, "none of the ordinary broadcast stations were usable because of lightning crashes from a thunderstorm somewhere over land to the west and north of Westhampton Beach." Furthermore, Hansell did not "remember observing any noise from lightning on the forty-one-megacycle signal from the Empire State Building even when ordinary amplitude modulation was being used." At this point the same newly outmoded theory of noise that Armstrong had used to invent wideband FM still held Hansell in its grip. He assumed that almost no static noise existed in the ultra-high frequencies. Yes, Armstrong's FM came through relatively clearly, but not, he guessed, because the system suppressed noise. Rather, Hansell insisted, "The elimination of atmospheric noises during the demonstration [of FM] was, I believe, due to the very low level of such noises on 42 [*sic*] megacycles as compared with noise on frequencies around 700 kilocycles [the standard AM broadcast band]." Moreover, widening an FM channel's frequency swing, he believed, amounted to asking for even more static; a wider deviation necessitated a greater "band width" [channel width], so double the swing, double the noise. Hansell calculated that "the noise at 7500 cycles should have increased about 10 to 1 in voltage when the transmitter wave deviated 75,000 cycles from the normal carrier value."[64]

Hansell chafed when Armstrong said (according to Hansell) that "he disagreed with AT&T engineers and others who had stated the noise output from a receiver is proportional to its selectivity band width. He said the noise was substantially independent of band width." In 1934 Armstrong's assertion—which we now know was correct—literally inverted what radio practitioners had regarded as a fundamental principle for at least ten years. Hansell especially doubted FM's ability to reduce static "where the noise is due to short peaks of very high intensity such as local lightning and inductive disturbances might produce."[65] Had Hansell listened under what today constitutes normal listening conditions, though, with a receiver placed at a short or medium distance from a far more powerful transmitter, and in an urban environment, he more likely would have realized that impulsive noises were precisely the kind that FM reduced best.

Hansell's reservations about the usefulness of FM fostered further misgivings in his mind about the strength of Armstrong's patent claims. "Armstrong has made a valuable contribution to the art of telephone transmission on frequencies above 30 megacycles," he admitted, and "his scheme will undoubtedly have wide commercial use." But the Armstrong system offered nothing new, according to Hansell, and "I am very doubtful of the validity of his patent claims covering the general principle of wide band modulation." "Probably only his specific circuit arrangements"—an allusion to Armstrong's balanced-modulator transmitter and balanced detector—"for which numerous alternatives are available, will stand up in court." Hansell buttressed this prediction by stating that RCAC engineers "used wide band frequency modulation at [RCA's] Rocky Point on Transmitter WQN, with a wave length of about 54 meters, in August 1925. The frequency modulation was applied to the transmitter continuously by the method described in my U.S. Patent 1,830,166, for reducing the effects of fading in telegraph transmission."[66]

Hansell could have cited abundant precedence for his doubts. Since the birth of broadcasting, noise-reduction schemes, virtually none of which had worked, had continually cropped up. Indeed, Armstrong himself had proposed several dead-end ideas based on the balanced amplifier, including, as mentioned earlier, a narrowband FM invention that he later recycled for the new system. Moreover, almost no evidence existed yet to back Armstrong's claims. Carefully controlled tests of wideband FM that produced quantifiable results still lay in the future, and in the absence of hard data, Hansell had little cause to alter his long-standing theoretical assumptions. After all, only a few months earlier the same assumptions had constrained Armstrong's thinking.

But Hansell also exaggerated the case against FM. Plainly, he fabricated his claim that RCAC engineers used "wide band" FM. A length of "54 meters" corresponded to a carrier frequency of approximately 5,500 kilocycles, a part of the spectrum where a 150-kilocycle frequency swing would cut an impractically wide swath. Nor did his Patent No. 1,830,166, or any other RCA patent, make a single wideband claim, as Hansell, who had filed thirteen other FM patent applications by 1934, must have known. Hansell also distorted the record when he attributed to "AT&T engineers"—an allusion to John Carson's 1922 article about modulation theory—the statement that "the noise output from a receiver is proportional to its selectivity band width." In fact, Carson's paper never mentioned wideband FM radio. Perhaps like Armstrong, Hansell misunderstood Carson's analysis.

Despite Hansell's skepticism, during the next several weeks Armstrong slowly accumulated a handful of allies within the company, principally by staging further demonstrations. These often amounted to little more than private performances, as when on 20 June he aired "a special program of organ music" for RCA's chairman, James Harbord.[67] He also found a more permanent and closer location for testing reception. A week later Armstrong drove Beverage, Hansell, and several other RCA engineers to the Haddonfield, New Jersey, home of Harry

Sadenwater to hear another demonstration of staticless FM. Sadenwater, a veteran RCA engineer and a longtime Radio Club of America member, lived near the RCA Victor factory in Camden, located fifty-six miles from the television lab. Armstrong had won him over as early as January, when, after he was "deeply impressed by the lack of audible interference from spark coil discharges [static]," Sadenwater and his wife permitted Armstrong to convert their basement bar into a more or less permanent test laboratory.[68] After hearing field tests of wideband FM for the first time in the Sadenwater home, Beverage reported that frequency modulation "was quite successful" in reducing static on the ultra-high frequencies. Even more impressive, he added, FM required a small fraction of the power of a standard broadcast station to achieve the same quality of sound. In fact, stated Beverage, AM stations in New York that radiated 50,000 watts sounded worse in Haddonfield than did the Empire State Building FM station, which used a mere 120 watts.[69] By the end of June 1934, Hansell's resistance to Armstrong FM had softened considerably. He admitted that the reception in Haddonfield of New York AM station WEAF on the evening of 25 June "was not particularly satisfactory and would not have been used for ordinary entertainment purposes," even though the weather "could have been far worse." "At the same time," he observed,

> wide band frequency modulation from the Empire State Building utilizing the same or similar NBC program was quite satisfactory and was generally free from noise and noticeable distortion. . . . It was quite evident that wide band frequency modulation transmitted from the Empire State Building on 41 megacycles with about 2 KW. power gave far more satisfactory reception at Mr. Sadenwater's house in Haddonfield than could be obtained over about the same distance from WEAF using about 50 KW. power on a frequency of 0.660 megacycles.[70]

These words marked an abrupt change of mind on Hansell's part. Less than two weeks earlier he had taken Beverage to task for saying much the same things, but now he estimated that because Armstrong's channel width could hold fifteen standard AM channels, the signal-to-noise ratio should improve proportionally.[71] This upended the theory he cited in Westhampton to justify his earlier skepticism. To be sure, Hansell remained wary about FM's commercial possibilities, and he continued to assume that operating in a nearly noise-free part of the spectrum caused the lower noise levels in the new system. But coming around to even part of Armstrong's theory so quickly speaks well for Hansell's intellectual flexibility.

Two months later, RCA further expanded the scope of its wideband FM tests.

The firm constructed two receivers, on the basis of the Armstrong patents, to assess long-distance FM. Beginning in October 1934, over a period of five months several engineers who lived in cities and towns located at least seventy miles from the Television Laboratory (chiefly in Pennsylvania and New Jersey) took this apparatus home to record reception data. Their weekly one- or two-page summary reports indicate that three central questions defined the goals of these tests:

1. How successfully does FM withstand interference of all kinds, including natural and man-made electromagnetic noises?
2. How does FM reception compare with AM with respect to noise suppression and maximum range?
3. How well does an FM signal propagate to receivers located at the extremities of reception; that is, a few miles over the horizon?

The narrowness of these questions, with their unfortunate overemphasis on "maximum range" goes far in explaining why, eventually, RCA declined to back wideband FM. RCA would typically test FM at long distances, beyond the range of normal reception, and with only a 2-kilowatt transmitter. The results of the tests added nothing to the understanding of frequency-modulation radio's most famous feature today, the ability to kill static, especially at shorter ranges, because, at the periphery of reception, signals sometimes came in strongly, at other times weakly. Moreover, no one determined any method to quantify answers to any of the central questions. Instead, observers wrote test reports in only qualitative terms. Charles Burrill, an engineer who listened to test signals in the third story of his home in Haddonfield, stated that that "during a portion of the time reception of frequency modulation was all that could be desired as regards background noise, while at other times interference which was probably quite local made reception impossible." He observed periods of fading that lasted up to an hour "nearly every day," adding that "apparently such fading is more likely at about noon or 6:00 p.m."[72] From another location "on the top of the Philadelphia Saving Fund Society Building," Burrill reported that ignition noise from automobile traffic "was so great as to make reception useless either for amplitude or frequency modulation."[73] From Milburn, New Jersey, he observed that "the noise level was very low except when an automobile was in the immediate vicinity. Consequently reception was very good with either frequency or amplitude modulation. Some of the time frequency modulation appeared to be somewhat better, but at other times there was no observable difference."[74] Though a capable engineer, Burrill's subjective observations, made at the "extremities of reception," hardly constituted replicable or easily analyzable data.

Why RCA Did Not Back Wideband FM Radio

Few articles of faith in the canonical history of FM radio resonate more with our sense of injustice than Lessing's explanation for why RCA opted not to buy the rights to Armstrong's FM patents. Lessing says that FM threatened to damage RCA's financial interests by upending the old order of AM radio. Frequency modulation, he asserts, "if allowed to develop unrestrained, posed a vast number of new radio stations, a complete reordering of radio power, a probable alignment of new networks, and the eventual overthrow of the carefully restricted AM system on which R.C.A. had grown to power."[75] Lessing also insists that RCA engineers, by and large, favored the development of FM but that they "could not overcome the weight of strategy devised by the sales, patent and legal offices to subdue" the threat of FM. Moreover, to emphasize the perfidy of RCA's managers, Lessing frames the company's neglect of FM in moral terms, as a "callous evasion of the most important new development in radio since the founding of the industry," and as an act of personal hypocrisy and betrayal on the part of RCA president David Sarnoff, who in early 1934 had "addressed a letter to [Armstrong] affectionately urging him to direct his energies toward the future of radio." RCA's subsequent refusal to develop FM, motivated by a fear of hurting the company's investments in AM, says Lessing, revealed "Sarnoff's [lack of] regard for that future."[76]

Several facts disprove Lessing's argument, though. First, it hangs on the false premise that RCA held substantial vested interests in AM broadcast radio technology. In fact, the company derived only a small percentage of its income from capital assets. In 1941 the FCC reported to Congress that, although RCA turned a profit every year since its inception, its capital investment, which "barely exceed $100,000,000," "would not be regarded as staggering" for such a large American company.[77] According to annual corporate reports, RCA earned far more income proportionally from such diverse noncapitalized sources as patent licensing, international communications, and the management of radio and recording artists. Furthermore, only a minuscule portion of RCA's total income came from the actual operation of AM radio stations. RCA owned the National Broadcasting Company, and although NBC provided programming to 111 network affiliate stations in 1935, the firm itself owned only ten outlets. Nine of these held lucrative clear-channel stations, but NBC's profits scarcely depended on the use of AM technology per se.[78] Rather, according to the FCC, NBC earned "about 90 percent" of its total revenues from 1926 to 1941 from selling air time to advertisers.[79]

Thus, neither RCA nor NBC had much to lose if FM displaced AM in American broadcasting.

Even the part of the RCA empire that relied on capital investment for income would gain little if anything by retarding the spread of FM radio. The FCC described RCA's manufacturing subsidiaries, such as RCA Victor and RCA Manufacturing, as "the largest single phase of RCA's business."[80] But the fact that the firm sold radio transmitters, receivers, and other apparatus did not translate into a motivation to preserve AM radio technology. On the contrary, the potential profit RCA stood to reap from the sales of new radio and television receivers significantly outstripped what the company might lose if it stopped making replacement parts for old AM transmitters.

Lessing's explanation also fails to account for why RCA, if it dreaded disruptions of the broadcasting industry, poured huge sums of money into developing television, on which David Sarnoff wagered RCA's future in hopes that the new technology would revolutionize American broadcasting. That RCA threw its full weight behind a technology that many hoped would make radiotelephonic broadcasting obsolete speaks to a corporate culture that welcomed, not feared, radical innovation, but only when sufficient evidence existed that convinced managers that a particular technology stood a good chance of working. If anyone in the company "opposed" wideband FM, he likely did so in the belief that Armstrong had not proved the value of his system.

Finally, the traditional explanation for RCA's inaction with FM makes sense only if RCA engineers and managers were fully aware of FM's features in 1935. But they could not have been. Armstrong himself cited only two unique advantages for his system over conventional AM systems: a greater range, which his patents mentioned and which went unproved, and a reduction of static noise, whose possibility the same patents implicitly denied. Moreover, by testing receivers so far from the underpowered Empire State Building transmitter, RCA engineers effectively obscured the desirable static-suppression properties of FM. Yes, some RCA engineers, like Beverage, Sadenwater, and Hansell, came around to suspecting or believing that wideband FM sometimes suppressed static. But these men could cite only impressionistic observations for their evidence, as no one had ever quantified the performance of the Armstrong system at ranges well short of distances where reception began to fade.

A far more plausible explanation for RCA's reluctance to buy Armstrong's patents is that company engineers and managers knew too little about wideband FM to assess accurately the potential of the Armstrong system. The flawed, nonquantified results of the long-distance field tests of 1934–35 largely account for this ignorance. Consider, for example, how poorly RCA engineers understood the relationship between transmitter wattage and reception. Armstrong later insisted that he had entreated RCA to lend him a more powerful transmitter, and if the company had complied, perhaps the static-reduction capability of FM would have become evident when the maximum range of the Empire State Building's signals extended beyond eighty miles. Rather than hearing the erratic reception that test engineers in New Jersey, Philadelphia, and Long Island continually reported, those men would have received something that resembled modern broadcast FM: essentially static-free radio, although with an audio bandwidth only marginally greater than that of AM radio. Moreover, though many RCA employees were unaware or unconvinced of FM's ability to kill static, even FM's most ardent advocates never speculated in 1935 that the Armstrong system could convey an audio bandwidth three times that of AM radio. This advantage, when coupled with low-static reception, would eventually make wideband FM the first high-fidelity mass medium by the end of the decade. But in 1935, no one knew of the feature, not even Armstrong, who in fact had not yet invented it.

The company's engineers wrote dozens of memoranda in that year reflecting these views, with none suggesting that RCA feared FM might end its AM-radio hegemony. In one typical evaluation, engineer Elmer Engstrom, who had observed several long-distance tests, stated in early 1935 that "near the transmitter the usual amplitude modulation system and the Armstrong system are both satisfactory. At intermediate distances from the transmitter, appreciable but not striking improvement is obtained with the Armstrong system. At greater distances from the transmitter but still within signal range neither system appears satisfactory in a practical sense although under some circumstances impressive gains have been noted with the Armstrong system." Engstrom did recommend that RCA develop an FM system "without excessive cost to R.C.A.," but he felt no urgency because the "Armstrong system is [merely] one application of the general principle" that "changes in the broadcast method of transmission are desirable from a technical viewpoint." RCA, he suggested, should keep involved with FM in case it becomes important. As for now, Engstrom declared, "the Armstrong patents do not appear particularly fundamental. They are unlikely sufficient to control the situation." Even if RCA were to acquire the Armstrong patents, "we could not operate without acquiring other patents held by outsiders."[81]

An even more dubious W. R. G. Baker, the chief engineer of RCA Victor, wrote to Otto S. Schairer, who headed RCA's patent department, that "while I would like to see RCA have a good patent position in this field, I can not see as yet the practical usefulness of the Armstrong system except for special services."[82] Ralph Beal stated to Schairer in May that the "predicted advantages" of FM "are its noise reduction characteristics and its ability to extend the transmitting range of ultra short wave broadcasting stations." Beal pointed out, though, that "measurements under average field conditions have not been completed but the results thus far have not shown gains altogether comparable to what may be predicted by theory."[83] A month later, Beal added that "the most important issue at this time with respect to the Armstrong system is the question of the extent to which the service area of a sound broadcasting station may be extended by its use."[84] These men were not obstructionists who feared a new technology's economic effect on the broadcast industry. Nor did they lack the imagination or heart to envisage a future high-fidelity broadcasting system. After all, many of them were working hard on developing the potentially far more disruptive technology of television. They merely had no evidence at hand that FM performed as its inventor claimed, beyond perhaps an impression that frequency modulation reduced static from time to time.

Taking FM to the Public

As RCA managers deliberated the fate of FM in the spring of 1935, Armstrong began to resent their lack of action. In April he signaled his impatience by enlisting journalists to plant the first seeds of FM radio's canonical history. Many of the most one-sided articles containing the most egregious and persistent myths about FM radio originally appeared in the newspapers of New York City. "Radio Device Ending Fading, Static Reported," proclaimed the *Herald Tribune* on 26 April. In addition to describing wideband FM in the most glowing terms, the newspaper alluded to Armstrong's fear of reliving courtroom disasters like his loss against de Forest over regeneration, by implying a lack of goodwill on the part of RCA. "Mindful of the legal arguments that may arise over a difference of a few months," the article stated, "Major Armstrong refused to say when he developed the new idea." The article asserted as well the false claim that all previous kinds of frequency modulation were narrowband FM. "The theory on which the new system works is a direct reversal of that on which engineers have previously worked to eliminate noises. The principle has been to narrow down the selective band as much as possible in order to keep down extraneous sounds, while the Armstrong system does just the opposite." Finally, Armstrong ignored the progress earlier researchers had achieved, especially at RCAC, by declaring that "the principle [of my new invention] is carried out by the use of a discarded method of modulation known as frequency modulation. . . . This method of modulation has been known for over twenty years, and the hitherto unsurmounted difficulties due to distortion and other troubles in both transmitter and receiver have caused its abandonment by all who worked with it." Armstrong did acknowledge that his Radio Club pals, Harry Sadenwater and George Burghard, had lent their homes for testing wideband FM, and that Charles Young, of RCA Victor, had experimented with multiplexed frequency modulation.[85] The *New York Times* said much the same thing, stating that FM was "nothing [Armstrong] merely stumbled upon. It is the result of long research in the laboratory at Columbia University and in the experimental station of the National Broadcasting Company atop the Empire State Building." But this nod toward the NBC lab was untypical, the *Times*' only hint that any company had assisted Armstrong in the development of FM. The newspaper also failed to mention the FM work of RCA and other companies that preceded the Empire State Building tests.[86]

Armstrong's decision to announce publicly the invention of wideband FM represented a major shift in his strategy for promoting his system. Earlier, he had consistently exalted the importance of secrecy. He told no outsider about RCA's FM research during the 1920s or about the Bolinas tests of 1931–33 and persuaded RCA employees to do the same. He concealed his own wideband FM investigations from RCA until his patents were securely in hand. He continued to shroud FM from the public for fifteen months afterward. Now, exasperated with RCA's indecisiveness, he turned to the press, partly to spur RCA to action, partly to vent his frustration.

After the initial announcements of FM appeared in the New York dailies, additional articles appeared in the papers of other cities, as well as in trade and professional magazines. In May, *Electronics* declared that "Major Armstrong seems to have solved the missing link in the battle against man-made and natural static."[87] The following month's issue said that Armstrong planned to provide a complete description of his system at the fall meeting of the New York section of the Institute of Radio Engineers. Although grumbling that Armstrong "has consistently refused to go into the details of his system," the magazine indicated that "several highly competent observers" of FM reception at Harry Sadenwater's home suggested that "no doubt . . . the system used actually does give a vastly better signal-to-noise ratio than conventional amplitude modulation methods."[88] *Communication and Broadcast Engineering* published a similar article, and many amateur radio operators first read about FM in the flagship magazine of the American Radio Relay League, *QST*, soon after Armstrong visited that organization's headquarters in New York City.[89]

For months, Armstrong's private publicity campaign elicited virtually no reaction from RCA managers. By midsummer of 1935, their taciturnity began to crack Armstrong's patience, imperiling his long-standing relationship with the company. About 1 August he staged still another demonstration in Haddonfield for two holdouts, Ralph Beal and David Sarnoff. A few days later he protested to Beal. "Mr. Sarnoff stated that his engineers wanted to make more measurements," he wrote.

> You will recall that I stated that I could not conceive how his engineers, after having been familiar with the system for over a year and a half and having witnessed numberless demonstrations, could require any further measurements to convince them of the utility of the system. As Mr. Sarnoff could not state what further tests were required, it was arranged that I meet Dr. Baker and yourself on July 29th, when you would advise me exactly what measurements were considered necessary.
>
> At this meeting neither Dr. Baker nor yourself informed me of any tests which you desired to make other than some very general statements that you wanted to find out "how far the system would work," a matter which I supposed had been thoroughly investigated during the tests of the past year and a half. Some "other tests" were referred to, but those present could furnish no information as to what they were. I therefore requested you, and you agreed to write me specifically, what further information it was that you required. To date I have received no word.
>
> In view of the fact that I have now been demonstrating [wideband FM] to the executive and engineers of the Radio Corporation under all conceivable conditions for a period of over a year and a half, may I now request you to advise me exactly what tests your engineers want to make and what the purpose of these tests may be.[90]

Armstrong neglected to state what he would do if RCA failed to heed his "request." In any case, on the same day that he composed this letter, W. R. G. Baker attempted to calm him down. "Interested groups in RCA," Baker wrote, are "making arrangements for certain tests on your system. One of the most important tests is to determine the effect of antenna height." Baker added that "we appreciate your offer to assist us in carrying on this work and would like your help in every phase of the program."[91] But this conciliatory gesture backfired. Baker's letter, the inventor shot back, "does not answer the question raised in my letter to you." Aside from the antenna height test, Armstrong continued, Baker made only "a vague reference to 'certain tests' which are to be arranged by you in connection with the 'interested groups.'" Armstrong explained that he had requested the antenna height tests more than a year earlier, even offering to install a frequency-modulation transmitter on top of Radio City, but that "this offer was not accepted." He now sharply demanded an unequivocal answer. "In view of the fact that my patents have been issued over a year and a half," he wrote, "or about 10% of their life, and the fact that throughout this time the executives and engineers have had full opportunity to apprise themselves of all phases of the situation, may I again ask you to state exactly what tests you want to have made and what the purpose of those tests may be."[92] This exchange signaled, if not the end, then certainly the beginning of the end of Armstrong's fifteen-year-old collaboration with RCA. Trials of the Armstrong system continued, but without the inventor's participation. Moreover, in October RCA began investigating the possibility of narrowing Armstrong's frequency swing, contravening the central principle of his system.[93] Inauspiciously, RCA disclosed none of this to Armstrong and a few months later he removed his FM apparatus from the Television Lab.

No one can ignore the ironies surrounding the development of modern FM radio. Its creator applied misconceived theories about frequency modulation and balanced amplifiers to invent a system that supposedly achieved little more than a reduction in tube hiss. Armstrong, in fact, implicitly denied in his patents that FM had any effect on static. Yet the same patents constituted the foundation of modern broadcast FM radio, whose static-reduction ability proved the wrongness of Armstrong's theories. Moreover, after 1936 FM would reveal additional advantages over AM, notably a wider audio bandwidth and greater resistance to interstation interference. But while the innovator who at first mistakenly denies that his brainchild does something useful might be unusual, he is hardly unique, particularly among independent inventors of complex systems. For years, Lee de Forest refused to acknowledge that the audion he patented in 1907 could amplify electrical waves, even though it is now famous as the first electronic amplifier.

What distinguishes Armstrong FM is the particular social and technological context in which it evolved as Armstrong and others uncovered the potential of the system. To understand how frequency modulation worked was not difficult in the abstract, but to make the first practical wideband system required first-rate practitioners like Armstrong or, alternatively, some of his colleagues at RCAC. Armstrong put into FM his long experience with balanced amplifiers and frequency modulation, both of which demanded far more stable circuits than AM technology commonly provided. Developing FM required as well substantial intellectual labor and material for testing. Up to a point, Armstrong provided these himself by constructing a prototype transmitter and receiver in his own laboratory and testing them over short distances. But unless he wanted to bear the far greater costs of constructing a transmitter with an elevated antenna, he would have to rely on RCA to demonstrate his chief original claim for FM: that it increased the range of ultra-high frequency transmissions. Armstrong would build such a station during the late 1930s, but only after the test facilities of RCA had enabled him to learn at little expense to himself that FM did far more than he had expected. For him to have done so earlier would have amounted to wagering illogically that his own preconceptions about FM radio were fundamentally wrongheaded.

Nor can one deny the role of extraordinary luck in the development of wide-band FM. Not one person during the 1930s predicted the full range of what the system would eventually accomplish. But serendipity in the Walpolean sense of the word more aptly describes the invention and discovery of the technology's static-reduction properties. Armstrong deserves credit for possessing the knowledge and open-mindedness—the sagacity—that allowed him to recognize that he had erred and that he had happened on something more useful than what he had originally anticipated.

FM Pioneers, RCA, and the Reshaping of Wideband FM Radio, 1935–1940

"Revolution in Radio"

Title of Fortune *magazine article about wideband FM, October 1939*

A much better title would have been "Civil War in Radio."

RCA engineer Ellison S. Purington, October 1939

In October 1935, more than five months after Howard Armstrong began leaking information about wideband FM to the press, RCA was still promising only more tests, which prompted him to escalate his offensive with a series of public demonstrations. This was an old strategy among radio practitioners. Guglielmo Marconi had taken his revolutionary wireless telegraph to the New York yacht races forty years earlier, and Lee de Forest staged several stunts to sell watered-down shares of his radio companies to gullible investors. Armstrong had done it himself fifteen years earlier. In 1921, as an officer of the Radio Club of America, he participated in the first transatlantic shortwave transmission, among the most widely publicized accomplishments in early twentieth-century telecommunications.

The first public demonstration of FM radio occurred before the New York section of the Institute of Radio Engineers, on the afternoon of 6 November 1935, in the old Engineering Societies Building located in Manhattan. Armstrong wrote to several RCA managers and engineers beforehand, suggesting that the firm use the occasion to announce also that RCA had transmitted facsimile images via FM from the Empire State Building to Haddonfield. As if to confirm Armstrong's suspicion that the company cared nothing about developing—and perhaps even feared—wideband FM, Charles Young, the RCA Victor engineer who was working with FM facsimile, coyly demurred: "We would not wish to issue any publicity which would detract from your paper." Armstrong replied that he planned to discuss Young's work anyway.[1] In any event, even if he could not parlay his relationship with RCA into broader publicity, Armstrong still hoped to change some minds in the organization, and he convinced several decision makers and influential engineers from RCA, including Murray Crosby, Harry Sadenwater, Harold Beverage, and Ralph Beal, to attend. In fact, Armstrong began his talk by thanking Sadenwater and his wife for the use of their basement, and acknowledging the assistance of several other RCA employees. Perhaps recognizing these individuals, coupled with the favorable impression FM would make on outsiders, might motivate insiders to take another look.

Almost all extant accounts of this demonstration indicate that many in the audience, especially journalists and others unaffiliated with RCA, witnessed for the first time in their lives the reception of a frequency-modulation program. Armstrong stood on a stage; behind him, he had arranged on a dozen or so tables the modules of a prototype FM receiver, interconnected with festoons of cables. The audience saw no FM transmitter, though, because the unit used that afternoon was located in Carman Runyon's house, twenty miles away in Yonkers. Witnesses recalled that Armstrong's longtime Radio Club friend Paul Godley assisted with the apparatus. He had also participated in the transatlantic shortwave demonstration in 1921; for the next several years, he would promote FM as an independent broadcaster.

Effective demonstrations of new technologies tend toward theatricality. Besides reading a draft of the paper that he submitted a few weeks later to the *Proceedings of the Institute of Radio Engineers*, Armstrong had Runyon transmit "staticless" speech from Yonkers. Then Armstrong played sound recordings RCA engineers had recently made on motion-picture film for the purpose of comparing FM with AM reception. For many in the audience, these samples left the greatest impression, because previously their knowledge about FM's superiority over AM came from only a smattering of newspaper and magazine articles. Even Ralph Beal, now RCA's director of research, and among the most skeptical engineer-managers in that company, admitted that "the presentation was especially interesting because of the [recorded-sound] demonstrations." He stated that they "were very effective in showing that the 41 megacycle Empire State channel was entirely free from static, whereas the present [AM] broadcast frequencies were practically blanketed."[2] A long article in *Electronics* similarly described "a very convincing demonstration of the new system. The quality of reproduction was as good as that of the best broadcast stations, and the interference level, produced by a noise-infested city area, was very low." A "dramatic demonstration," echoed *Communication and Broadcast Engineering*.[3]

The most famous—and by far the most inaccurate—account of Armstrong's performance was written by a man who was not there. Lawrence Lessing claims in *Man of High Fidelity* that the audience heard demonstrations of sound effects that had so far thwarted the best technology available in broadcasting. Armstrong, Lessing writes in 1956, treated his audience to an exhibition of what became "part of the Major's standard repertoire in showing off the remarkable properties of his new broadcasting system":

> A glass of water was poured before the microphone in Yonkers; it sounded like a glass of water being poured and not, as in the "sound effects" on ordinary radio, like a waterfall. A paper was crumpled and torn; it sounded like paper and not like a crackling forest fire. An oriental gong was softly struck and its overtones hung shimmering in the meeting hall's arrested air. Sousa marches were played from records and a piano solo and guitar number were performed by local talent in the Runyon living-room. The music was projected with a "liveness" rarely if ever heard before from a radio "music box." The absence of background noise and the lack of distortion in FM circuits made music stand out against the velvety silence with a presence that was something new in auditory experiences.[4]

Doubtless such an exhibition of realism would have dazzled listeners in 1935, but Lessing's description is almost entirely fictional—or at least premature. In fact, the preceding quote amounted to a paraphrasing of newspaper and magazine articles written about demonstrations of "high-fidelity" FM two years later. No one who attended that day reported hearing reproductions—vivid or otherwise—of crumpled paper, oriental gongs, guitars, pianos, or Sousa marches. Nor could they have, for Armstrong had yet to incorporate the "high-fidelity" circuits into his system that the reproduction of such sound effects requires. In 1935 even live, point-to-point wired high-fidelity sound reproduction remained principally the esoteric hobby of a handful of audiophiles.

It is true that Armstrong was hoping someday to incorporate circuits into his system that would, ultimately, dramatically widen the audio bandwidth of FM. *Communications and Broadcast Engineering* reported that "Professor Armstrong . . . pointed out that due to 'the extremely short wavelengths, it has been possible to transmit all modulation [audio] frequencies from 30 to 16,000 cycles, and to receive them with what engineers call a flat characteristic [i.e., zero distortion],'" a statement that closely comports with the 15,000-cycle flat response of modern

FM.[5] And one of the two wideband FM patents of 1933 had declared that FM was suitable for television or facsimile "where the rates of modulation are much higher than in voice transmission."[6] Further, an NBC engineer recalled fifty years later that in December 1934 NBC widened the audio bandwidth of wired circuits between Radio City and the Empire State Building laboratory from 10,000 to 14,000 cps, "undoubtedly," he asserted, "part of Armstrong's plan to establish and demonstrate FM as a very-high-fidelity system."[7] But no evidence exists that Armstrong used such circuits until 1936, and no one who attended the IRE demonstration reported hearing sound effects that would have required those circuits. At best the audience was served up a sound quality that far exceeded AM's performance in terms of static suppression, but only moderately, if at all, in terms of audio bandwidth—comparable to the audio fidelity of a late-twentieth-century telephone.

Lessing's 1956 account of the talk, however spurious, is nonetheless notable as a historical artifact, for it echoes the strategy to sell FM outside RCA that Armstrong began to piece together after the New York demonstration. Persuading others of the advantages of FM in 1935 presented a daunting challenge. As the largest radio company in the world, RCA commanded unrivaled respect, and Armstrong needed to quash suspicions that the firm's engineers had rejected FM on technological grounds. In fact, RCA did spurn Armstrong FM for technological reasons. That is, too many engineers and managers—Ralph Beal among them—underrated the true potential of the Armstrong system, a colossal error on their part due chiefly to a combination of intellectual inertia and inadequate testing. The company would never admit to such incompetence though, and its prestige was great enough to blunt the plausibility of the truth. Therefore, Armstrong began to assert that RCA acted out of malice and fear and, moreover, that the firm not only declined to back but also opposed FM radio.

At the end of the day, all the RCA men who came to the IRE talk in New York unconvinced left unconvinced. Even Ralph Beal, who had raved over RCA FM's triumph in the 1931 Schmeling-Stribling fight overseas broadcast, now rejected Armstrong FM. His reasons rested, regrettably, on old-fashioned misapprehensions about the relationship between static and the spectrum. Beal reported to Otto Schairer that the quieter reception did not result from using FM instead of AM, but rather because wideband FM operated in the ultra frequencies instead of the noisier parts of the spectrum. "Major Armstrong," he said in making this point, "did not comment on the fact that . . . there is practically no static on the ultra short waves." Beal also implied hucksterism on Armstrong's part, accusing the inventor of using a "considerable display of showmanship for the purpose of putting over the idea of the wide band system." "In this respect," he told Otto Schairer, who headed RCA's patent department, "I feel that [Armstrong's talk] deviated from a conservative report of a new development to a body of competent engineers."[8]

The IRE demonstration marked several milestones in the history of FM radio technology. First, it was Armstrong's last failed attempt to sell his patents to RCA. Second, the talk signaled the point at which FM ceased being the concern of a single organization. No longer would RCA be the only important locus of FM research. By taking his system to the public, Armstrong had ensured that from now on, his invention would be developed by a wider community of broadcasters, engineers, and corporations, who took to calling themselves "FM pioneers."

By far the two most significant individuals to enroll in this community were John Shepard 3rd, and Paul DeMars, the owner and chief engineer, respectively, of the New England–based Yankee Network. Shepard, one of the greatest broadcast entrepreneurs of the twentieth century, had been elected the first vice president of the National Association of Broadcasters in 1923. During the same year he invented a crucial element of the American system of broadcasting—the network—by leasing long-distance telephone lines from AT&T for the purpose of simultaneously duplicating live programs in remote cities.[9] Radio networks adopted the technique during the late twenties, and it fell out of usage only after World War II, when microwave radio relays (using frequency modulation) replaced wires.

How the Federal Communications Commission regulated AM broadcasting accounted for why Shepard embraced FM. The commission had crafted a policy of extending radio service to remote, usually rural areas by creating an intricate hierarchy of stations. At the bottom were hundreds of short-range, low-power stations with, 250-, 500-, and 1,000-watt transmitters. These operated only during daylight hours when radio waves propagated over relatively short distances. Above them was a smaller group of various classes of stations that broadcast for longer periods of time and with greater power—5, 10, and 50 kilowatts, for example. At the top were privileged, usually 50-kilowatt, twenty-four-hour stations that occupied one of a few dozen "clear channels" that no other broadcaster used, and which interstation interference, therefore, only minimally afflicted. This three-tier arrangement significantly reduced congestion at night and made network programs available to all but a few corners of the country. But by allowing only a small number of chiefly high-power stations to broadcast at night, the FCC effectively relegated most stations to second-, third-, or fourth-class status.

John Shepard represented a large group of broadcasters who resented clear channels, for primarily economic reasons. A commercial station made money from airing commercials, and the larger its audience, the more a station could charge for advertising. How many potential listeners a station reached depended in turn on a matrix of factors, including listener ratings and network affiliation, but the most important by far was how high a station ranked in the FCC's hierarchy. The principal station of Shepard's Yankee Network held a regional license, which permitted 50 kilowatts of power, the legal maximum, but like all regionals, it shared its channel with other stations, so that when the radiation patterns of another station on the same channel overlapped, the programs of both were ruined. Because this kind of interference never affected a clear-channel broadcast, virtually all regional stations earned substantially less revenue than a clear-channel station. To make matters worse for Shepard, none of the handful of clear-channel broadcasters in New England was likely to give up its license voluntarily. This state of affairs drove him to political activism. In 1938 Shepard was elected the first president of the National Association of Regional Broadcast Stations, an organization that supported a policy of minimizing the number of clear-channel licensees.[10]

FM provided non-clear-channel broadcasters like Shepard the hope of rendering wattage and clear channels all but irrelevant. Because the range of even a high-power, high-frequency station was limited to a few dozen miles beyond the horizon—day or night—all FM stations were local ones. To be sure, as a regional AM broadcaster, Shepard could not claim that the FCC was grievously wronging him by refusing to grant his chain a clear-channel license, but early on he realized what many progressive critics of radio later understood about Armstrong's system: that by offering a technological fix that obviated the FCC's hierarchal system, the new medium might democratize the broadcast industry and revive local and regional radio. Nonprofit broadcasters, largely comprising a small number of educational stations that had survived an earlier weeding-out process by the FRC, also stood to gain, because FM also allowed for more stations to be on the air.[11] FM could be, as more than one writer put it during the forties, "radio's second chance."[12]

The Yankee Network's association with FM radio began to form perhaps as early as 1935. Armstrong won Paul DeMars over straightaway—only a few weeks after the IRE demonstration—and DeMars in turn brought John Shepard into the camp. Positive results for FM followed almost immediately, as in April 1936, when DeMars helped Armstrong persuade the FCC to allocate to wideband FM an experimental portion of the radio spectrum. The commission set aside 42.5 to 43.5 megacycles and 117 to 118 megacycles, enough for ten channels (although for technical reasons only the lower-frequency band initially proved useful for broadcasting).[13]

The importance of FM pioneers like Shepard and DeMars in accelerating the social and technological evolution of FM radio cannot be exaggerated. In 1935 only two active wideband FM stations besides the Empire State transmitter existed: Armstrong's, located in his laboratory on the campus of Columbia University, and Carman Runyon's small rig in Yonkers. By the close of 1937 Armstrong had begun to construct a 40-kilowatt station in Alpine, New Jersey, across the Hudson River from Manhattan. Shepard and DeMars were also building a 50-kilowatt station for the Yankee Network in New England, and Paul Godley planned to operate a low-wattage transmitter in New Jersey.[14] Two other early enlistees were Franklin M. Doolittle and Daniel E. Noble, both electrical engineers. (Doolittle owned AM station WDRC in Connecticut.)[15] One especially distinguished pioneer was John Hogan, the engineer of an experimental high-fidelity AM station in New York City, W2XR. A historical figure in his own right, Hogan had begun his career assisting Fessenden and de Forest thirty years earlier in their groundbreaking amplitude-modulation radiotelephony efforts. Now Hogan himself was soldiering in another revolution. In 1939, after hearing the Armstrong system, he converted W2XR into what became WQXR, the first regularly scheduled FM broadcast station in Manhattan. By 1940 dozens more had joined him, including the owners of several regional AM stations, and large radio apparatus manufacturers such as Stromberg-Carlson, General Electric, Radio Engineering Laboratories, and Westinghouse.[16]

Armstrong also stepped into the ranks of FM pioneers. In April 1936 he became an independent broadcaster, when he announced plans to build an experimental "high-power" FM transmitter, and in June the FCC approved his request to begin construction.[17] Even for a multimillionaire like Armstrong, his expenditures, made at the midpoint of the Great Depression, represented a courageous personal commitment to the future of FM broadcasting. On his application form to the FCC, he estimated the transmitter's cost at $48,000, and other apparatus at an additional $9,000.[18] Eventually, at least $250,000 went toward a transmitter building and a spectacular 400-foot tower with a trio of 150-foot crossarms.[19] But those sums covered only the down payment. Four years later Armstrong admitted that he had spent between $700,000 and $800,000 of his own funds on FM, including $300,000 for his station.[20]

Evolving toward Modern FM Radio

During the late 1930s, Armstrong and other FM pioneers continued to realize new advantages for wideband FM. First, Armstrong began to improve the overall fidelity of FM by incorporating circuits capable of reproducing sound with audio frequencies up to of 15,000 cps. He never explained how he decided on this standard, but most likely the idea grew on him as the practice of reproducing high audio frequencies became easier with experience, and as he apprehended the resulting manifest improvement in sound quality. He seems to have spread the word about this innovation much as he had about FM's static suppression properties in 1934: he told friends and staged demonstrations that showed that FM's fidelity had improved to such a level as to match Lawrence Lessing's descriptions that he incorrectly attributed to the 1935 IRE presentation. On 16 March 1938 Harry Sadenwater declared to his supervisor at RCA Manufacturing that FM now "allows the full audible range of sounds from thirty cycles to seventeen thousand cycles to be transmitted without noise or hiss in the program. And the difference in naturalness of reproduction is actually startling. I have never heard quality that would equal that demonstrated over Armstrong's apparatus."[21] Five days later the inventor showed off his newly developed high-fidelity FM system at a Radio Club of America meeting. *Broadcasting* reported that audience was "visibly impressed with the clarity and freedom from noise. The sounds of tearing paper, pouring water and ringing bells and chimes might have been coming from the … same room as far as the ears could detect."[22] In May, Armstrong presented a similar demonstration in Boston. Henry Lane, the technical editor of the *Sunday Post* called the event "the largest gathering of broadcasting and communications engineers and scientists ever to meet in Boston under the auspices of the Institute of Radio Engineers":

> Virtually spellbound, nearly 600 college professors, engineers, technicians, scientists and the curious sat for well over an hour listening to all manner of programme material, including vocal, instrumental organ, band and orchestral music, together with sounds such as tearing of paper, the pouring of water and the striking of a bell. These things were heard as they have never before been heard over a radio system. Not the least impressive feature of the new system is the practical absence of any form of background noise. Static, tube hiss, hum and the other distracting sounds that nearly always accompany conventional radio reception is entirely lacking.[23]

"The system," declared Lane, "will reproduce silence itself."[24] Such fidelity was, for all practical purposes, equivalent to what monophonic FM accomplishes routinely today. With the 15,000-cycle bandwidth now normal practice, coupled with wideband FM's already-well-known static reduction capability, FM in 1939 was the first truly low-noise, high-fidelity mass medium, setting a standard that other audio technologies, such as motion-picture sound, long-playing phonographs, and magnetic tape recording, never matched until after World War II.

In 1939 engineers at the General Electric Company announced the discovery of yet another surprising advantage over AM radio—namely, that wideband FM almost completely suppressed interstation interference. Listeners who tuned to an AM channel occupied by two stations simultaneously heard the garbled chatter of both programs combined. GE engineers had found that, by contrast, only the stronger of two FM signals was audible under comparable conditions.[25] So abruptly does an FM receiver switch from one signal to another that when a radio was installed in an automobile located at the point where the signals were approximately equal, "the movement of the car a few inches was enough to change the signal from one station to the other[,] and at practically no point were the observers able to get both signals simultaneously."[26] The implications of this news were extremely propitious for a future national broadcasting system. Now that GE had proved that interstation interference affected FM dramatically less than AM, the FCC could place FM stations much closer to each other both geographically and on the radiofrequency spectrum, accounting for a famous paradox of FM: although an individual FM channel spanned twenty times the spectrum of an individual AM channel, far more FM stations than expected could operate simultaneously without "crosstalk" in a large region. In other words, the properties of wideband FM had the potential effect of conserving spectrum generally, which hastened, as much as anything did, the FCC's acceptance of the system.

The most spectacular demonstrations during the late 1930s and 1940 proved the feasibility of wideband FM networks. In January 1938 Shepard and Armstrong announced their joint investment of half a million dollars in a network of relay stations that would allow for "catapulting . . . [radio program] signals from substantial heights" and over long distances.[27] The object of the project was to lay the foundation for a new technology of chain broadcasting, to replace Shepard's 1923 method for AM-radio networking. Under the old system, CBS and NBC leased telephone lines to transmit live broadcasts across the country from a studio, but wire lines carried no more than 4 or 5 kilocycles in audio bandwidth, only one-third of FM's capacity.[28] Shepard, DeMars, and Armstrong therefore proposed transmitting point-to-point full-channel programs from one high-altitude station to another via FM on 200-kilocycle-wide channels in the 110-megacycle band, and then broadcasting the programs locally on the 42-megacycle band.

Two years later they did so with a pair of "triple-play" relays. On 3 December 1939 and 4 January 1940—the latter date chosen because it marked the seventeenth anniversary of Shepard's chain-broadcasting idea for AM radio—Carman Runyon beamed sixty-minute programs from his home in Yonkers to Armstrong's Alpine tower. Armstrong then relayed the signal eighty-five miles to the Connecticut mountaintop antenna of W1XPW, Doolittle and Noble's experimental FM station. Finally W1XPW passed the signal on to W1XOJ, Shepard's station in Paxton, Massachusetts, which broadcast the program to metropolitan Boston.

To hear any audible speech or music after three relays would have astounded radio engineers, because even one or two comparable legs on AM distort the signal intolerably. But the quality of reception far surpassed even the most sanguine expectations. Henry Lane of the *Sunday Post* reported that "the program itself was designed to subject the system to a severe test for quietness and fidelity. Selections by piano, guitar, violin and brass instruments singly and in combination, high grade transcriptions and special sound effects served to give the listener an amazing demonstration." Lane added: "On top of this, the quality of reception in Boston with the nearest transmitter 45 miles away was fully up to a direct broadcast and showed no apparent loss of quality. Quite evidently, the process of rebroadcasting can be carried to a point far beyond that used in this initial test. The quality? You must hear it to understand how good it is. 'Natural' is the best descriptive word."[29] K. B. Warner, the longtime editor of *QST*, agreed, declaring that "it was just technically unbelievable with three relays, yet the program was still better by far than the present conventional [AM] system at its best." "In 10 years," he predicted, "there won't be any orthodox brand of broadcasting [AM radio] remaining except for the lowest grade of local service."[30]

Buoyed by this triumph, Shepard turned to the political side of FM, and on the day after the 4 January demonstration convoked in New York "73 individuals representing 49 organizations" to charter FM Broadcasters, Incorporated (FMBI), an organization dedicated to promoting wideband FM. *Broadcasting* reported that, of FMBI's members, "12 already have F-M stations: 10 have construction permits; 22 have applied for construction permits and nine propose to file such applications before . . . Feb. 28." On the founding committee sat representatives of several large regional AM stations, as well as engineers from Stromberg-Carlson, General Electric, Scott Radio, and the Radio Engineering Laboratory, a company that worked closely with Armstrong in the manufacture of FM transmitters. Even RCA and NBC sent representatives.[31]

Despite the fact that Armstrong's system had spawned a diverse community of FM pioneers, only a few squabbles impaired the harmony of the first FMBI meeting. The most serious debate turned on the question of whether FM should be allocated a band immediately above 44 megacycles, where the FCC currently assigned television's Channel 1. Television developers understandably said no. A representative from Zenith Radio "suggested that F-M stay away from television and confine its activities to the frequencies around 100 mc." Also, Armstrong clashed with NBC's O. B. Hanson about the necessity of 200-kilocycle-wide FM channels, a standard that FM practitioners had long made permanent and that exists today. Hanson asserted that a narrower channel might suffice, and he promised that after RCA completed a series of forthcoming tests that he would provide proof—evidence that never materialized.[32]

These disputes amounted to minor distractions, though. In addressing its more important concerns, the group unanimously passed a resolution asking the FCC to begin issuing "regular," not merely experimental, licenses to frequency modulation stations; to increase the number of FM channels from five to fifteen; and to locate the future FM band near the current one of 42.5–43.5 megacycles, preferably from 41 to 44 megacycles.[33] Indeed, Shepard had already paved the way for these proposals. In October he had petitioned the FCC for a hearing to grant the Yankee chain "a regular license as distinguished from an experimental license." This amounted to a call for the FCC to issue commercial licenses on a routine basis, because restricting the privilege to a single station would make little sense. In early December, E. K. Jett, the FCC's chief engineer, met with Armstrong and subsequently ordered a "study which will compare F-M potentialities with amplitude modulation."[34] Finally, on 19 December the FCC announced that Yankee would get its hearing.[35] After polling dozens of FMBI members and other interested parties, the panel scheduled a date: 18 March 1940.

RCA and FM Radio during the Late 1930s

The presence of RCA representatives at FMBI's inaugural meeting raises the question of what RCA had done with frequency modulation since passing on the Armstrong system four years earlier. The answer is, even to be kind, not much, aside from articles about phase and frequency modulation that Murray Crosby had published in the *Proceedings of the Institute of Radio Engineers* and the *RCA Review*.[36] Tellingly, RCA acquired the rights to no FM patents during the years 1937, 1938, and 1939, precisely when Armstrong and others were hammering out the specifications of modern broadcast FM radio.

One sign that Armstrong intended to isolate RCA from FM was his expensive decision in 1937 to ask General Electric, not RCA, to build the first batch of FM tabletop receivers. In May, GE quoted $900 for one set, but Armstrong negotiated a lower unit price by agreeing to buy twenty-five units at around $400 each.[37] (He would typically present these receivers to FCC commissioners, members of Congress, and other individuals with the power to influence FM's future.) Contracting General Electric to assemble a few FM receivers proved costly, but he could not abide doing business with RCA, which might have built cheaper sets. When Harry Sadenwater inquired "why [RCA Manufacturing] had not been given an opportunity to make these receivers," Armstrong replied, according to Sadenwater, "Because the RCA Patent Attorneys were trying to steal [my] invention."[38]

In fact, most activity within the company centered not at all on thievery but rather on a pointless, mostly after-the-fact debate about FM's commercial feasibility. Determined doubters like Ralph Beal at first felt no qualms about letting Armstrong's system go, chiefly because they simply questioned whether frequency modulation could reduce static noise, resolutely holding to the conviction that wideband FM's lack of static resulted only from the fact that the system operated in the ultra-high frequencies. Armstrong's claims for FM, Beal still insisted at the end of 1937, amounted only to "coupl[ing] with his modulation method the advantages of freedom from atmospheric disturbances and ability to obtain better quality by the use of a greater channel width." "Both of these advantages," explained Beal, "are common to any ultra short wave station regardless of the modulation method employed."[39]

Supporters of Armstrong FM within RCA—all engineers, and none of them senior managers—comprised a tiny, often cautious faction. Clarence Hansell, for instance, wished Armstrong success with the station he proposed to build in Alpine: "I have been trying to find an opportunity for a number of years to get frequency modulation transmission into commercial service." He cordially promised that he "will be watching with considerable interest your efforts to establish a frequency modulation broadcast transmitter."[40] Harold Beverage also believed in wideband FM, but he had all but given up on any fight to bring RCA around.

In contrast to Beverage and Hansell, who despaired of recapturing RCA's lead in the FM race, Harry Sadenwater stalwartly advocated frequency modulation—more than anyone else in the organization did. Although only a rank-and-file engineer, Sadenwater bravely took to scolding his superiors in the company for neglecting wideband FM. Predicting "approximately 1,000 50 kw. [FM] stations" in the near future, he recommended to a vice president in early 1936 that "our advanced development group begin to outline commercial [FM] apparatus to meet possible inquiries from our customers."[41] Two years later, upon learning that NBC proposed to spend $8,500 on an experimental high-fidelity stereo *amplitude-modulation* radio experiment, Sadenwater barely contained his outrage as he explained that FM had already produced high-fidelity and ultra-high-frequency broadcasts. "From almost every technical viewpoint that I can visualize, the ultra-high frequency broadcasting that is developing will undoubtedly finally utilize frequency modulation," he asserted. Rather than squander money on high-fidelity AM radio, Sadenwater stated, "we should be prepared to supply frequency modulation equipment."[42]

On 16 March 1938, soon after receiving his first pay raise in almost a decade, Sadenwater dispatched an especially pointed memorandum to C. K. Throckmorton, the executive vice president of RCAM. The letter conveyed the tone of an aging ("grown gray," as Sadenwater described himself) company man who had paid his dues and now demanded to be heard. He recounted how he had operated an amateur station in 1908, "when antennas were far and few between." He "taught radio school in N.Y.C." from 1914 to 1917, and as a lieutenant (junior grade) had served as radio officer on NC-1, one of the group of four navy flying boats that attempted to cross the Atlantic in 1919 (Sadenwater's craft made a forced landing short of the Azores, but one of the other airplanes completed the trip). As a field engineer for General Electric during the twenties, he built several high-power broadcast stations before transferring to RCA in 1930.[43]

In Sadenwater's opinion, RCA had fumbled FM, "the most important subject for any of us in the radio manufacturing business." "Frankly," he admitted, "I've been a bit discouraged because it's been a long, long time since any increase in salary has come my way. . . . But because I have faith in radio and the fundamentally sound need for RCA I have repeatedly determined to stick to the ship." Sadenwater led Throckmorton point-by-point through the case for Armstrong's system. Allowing that AM receivers had saturated the consumer market, he nonetheless insisted that the public would buy more radios if RCA were to offer "something new and appreciably better." He declared that "the sounds made by [AM] broadcast receivers and motion picture reproducing systems are horribly distorted. It really makes me irritable and nervous to listen to them." He stated that "good fidelity in the existing [AM] broadcast band is impossible due to the large number of stations and the few channels available." "From every angle that I have looked at the new [wideband FM] system, my conclusion has been favorable to it and I am sure it will ultimately be the system used for broadcasting."[44]

Sadenwater blamed some of the most powerful men in RCA for the firm's mistakes. "I have discussed this matter with our engineers," he told Throckmorton, "and Mr. Clement [an RCAM vice president] seems to be positive in his conclusion that it is of no importance. I can only believe that he does not know enough about it and that the reports on which he had based his conclusions were not well founded on good data and on enough experience." Sadenwater questioned the judgment of one engineer-manager in particular, a man who had participated in KDKA's FM experiments of the 1920s. "I have also many times discussed this matter with C. W. Horn, the Development Engineer of the N.B.C. Horn says it is impractical to hope to replace the great quantity of receivers now in the hands of the public, representing an investment of several billion of dollars. My answer is that it has been done, gradually, twice before and will be done again. Horn also questions the practicality of discontinuing the present broadcast service being rendered by [AM] stations. . . . As I see it, it could be worked out with time."[45]

As an overt champion of FM, Sadenwater stood virtually alone in RCA until the end of 1938, when he obtained an ally in Dale Pollack, a fellow RCAM engineer and a recent graduate of the Massachusetts Institute of Technology. Initially, Pollack numbered among the skeptics. In late 1937, he had gingerly suggested that RCA merely keep a hand in frequency modulation. Predicting "considerable application of frequency modulation in the near future," he recommended only "a new method of frequency modulation, simpler than Armstrong's."[46] Fourteen months later, though, Pollack began to agree with Sadenwater, after hearing a talk that W. R. G. Baker, the manager of General Electric's Radio and Television Department, gave on FM at an IRE meeting in Rochester. Indeed, Baker himself had recently converted from skeptic to crusader in the army of FM pioneers. From 1929 to 1935 he had been the production manager of RCAM, where he consistently weighed in against the development of frequency modulation.[47] Since joining General Electric, however, he had shifted to the other side of the issue and, with the conviction of a repentant sinner, was earning a reputation as the driving force behind GE's support of Armstrong FM. Pollack reported to his superiors that "from the tenor of [Baker's] introductory and closing remarks it was evident that the advantages of frequency modulation are fully appreciated by General Electric. . . . Some of Baker's remarks on this point were quite emphatic." Pollack concluded that "the principal thing impressed upon me . . . is that a great deal of work has already been done [by RCA's competitors] on frequency modulation. . . . If we are not to be left behind our development should be accelerated."[48]

Pollack urged that RCA rectify "three broad problems . . . if we are to learn to design frequency modulated transmitters": the firm's lack of "practical circuits for producing frequency modulation"; the need to design measuring equipment to assess the performance of the not-yet-designed transmitters and receivers; and, finally, an institutional ignorance in the field of frequency-modulation theory.[49] RCA, once the cynosure of FM research and development, had lost, in Pollack's opinion, almost all the often-tacit knowledge necessary to design practical apparatus. Aside from Murray Crosby's important theoretical work, Pollack was right. RCA had done almost nothing with FM since letting the Armstrong system slip its grasp and had almost forgotten what it learned during the previous decade of research. The firm that prided itself as leading the vanguard in telecommunications research could barely manage to bring up the rear of FM development.

Conservatives who opposed the development of FM eventually realized their error, however slowly—but not because of the technical reasons that Sadenwater and Pollack had pointed out. The widely publicized achievements of such FM pioneers as Shepard, DeMars, and Armstrong himself carried far more weight. In late 1938, after a flurry of press releases and articles lauding FM issued from the popular, engineering, and broadcast industry press, Ralph Beal scheduled a staff meeting for 20 January 1939, two weeks after Armstrong and Shepard's second mountaintop relay demonstration and the FMBI's first meeting, "to consider the subject."[50] Dale Pollack (but not Harry Sadenwater) sat in, and afterward Pollack wrote his bluntest criticism of RCA's policy. Because RCA lagged so far behind, he insisted, its engineers should organize "an intensive development program with as little delay as possible." Pollack added that he had reviewed the Empire State Building FM test reports of 1934–35 and had made a disturbing discovery—namely, that the trials had been so badly planned and carried out as to render their results worthless. He included among the dozen flaws he listed that "many more listeners under a wider variety of circumstances should have been used, and more measurements should have been made. Only one listener at a given location was used and the period covered, two to seven days at each location, was too short." He also cited the low transmitter power used at the time—only 2 kilowatts, in contrast to the 40 and 50 kilowatts that Armstrong and Shepard were now using successfully.[51] Again, Pollack was right; moreover, the early tests evaluated FM performance chiefly at the periphery of its radiation pattern, which effectively guaranteed erratic reception.

Even when RCA managers began, in late 1939, to accept the reality of wideband FM, an institutional arrogance about the Armstrong system retarded their transformation. Some employees believed that FM, even at this late date, could not survive economically without RCA's backing. Clarence Hansell, for example, told Niles Trammel, the new president of NBC, that "the Major and other investigators cannot put frequency modulation over on a commercial basis in broadcasting unless a company like RCA, controlling a manufacturing company and a broadcasting chain, sponsors it."[52] O. B. Hanson also brushed aside the FM pioneers, who, he believed, foolishly presumed to displace RCA from its role as a telecommunications leader. "It is doubtful," he stated in a memorandum written four days after Beal's meeting, "if individual investigators who are now building frequency modulation transmitters can, by themselves, swing the industry in that direction." He continued: "Whatever system [of FM] is adopted by RCA, its manufacturing company and its broadcasting company, will probably be the governing factors in the future."[53] The same assumptions underlay Hanson's proposal, at the inaugural meeting of FMBI a year later, to adopt a standard channel width narrower than 200 kilocycles per second. Incredibly, he made this suggestion despite the fact that RCA engineers had logged almost no practical experience with FM technology for half a decade.

This smugness arose in large part from a universal belief within the company that FM posed no economic threat to RCA. In early 1939, Hanson explained that whatever he disliked about frequency modulation, he did not fear the injury the new system might inflict on RCA's investment in AM radio. "Regardless of what technical system is used," he assured the president of NBC, "the expansion of broadcasting in the ultra short wave field will have its effect on our company by the dilution of the listening market. This, in itself, is not too serious in my opinion, as in the last analysis it is the program material that gets the listeners."[54] In other words, the profitability of RCA, through its subsidiary company NBC, rested more on whether listeners tuned into the network's radio programs than on the kind of technology that carried those programs.

RCA's managers did, to be sure, eventually come to exhibit nervousness about FM, but only after realizing that their proud company might fall even further behind its competitors, especially Westinghouse, Stromberg-Carlson, Zenith, and General Electric. Armstrong had already licensed at least nine companies that attended FMBI's first meeting to use his patents to manufacture FM receivers.[55] "All indications," warned one RCA manager in May, 1939, "are that [General Electric and Westinghouse] are going to promote frequency modulation, and their activities, together with those of the REL, are making it daily more embarrassing for us and I therefore feel that we must get ourselves in a position to be able to furnish quotations to broadcasters on frequency modulation transmitters and receivers."[56] "CBS," Clarence Hansell informed Niles Trammel, "is filing [an application with the FCC] for a channel for frequency modulation. We should do the same."[57] Trammel's response exemplified how much apathy—an apathy anchored in ignorance and complacence—still pervaded the feelings of NBC's managers about FM: "I hope you keep me advised of the developments as they occur. . . . Should CBS engage in frequency modulation, how much of an advantage will it give them over us?"[58]

By the spring of 1939, RCA had sloughed off a few layers of this indifference, with several managers acknowledging that FM pioneers had transformed a technology that once seemed of dubious value into one that now appeared inevitable. Ralph Beal signed off on an engineering report that recommended "the adoption and use of frequency modulation for the transmission of sound in all domestic broadcasting services which use ultra short waves." But the company's managers were also determined to leave RCA's stamp on the medium before the FCC "black-boxed" it—that is, before the commission established permanent technical standards. Although every FM practitioner outside RCA had accepted Armstrong's 150-kilocycle swing as normal practice for more than six years, the Beal report declared that "no conclusion was reached [by RCA engineers] as to the amount of deviation or frequency swing to be suggested."[59]

Another indication of RCA's inability to grasp its FM problem was that, because of the firm's paltry recent experience with frequency modulation, company engineers could not build practical, commercial-quality, apparatus. In June 1939, R. D. Duncan Jr. of RCA Manufacturing's Transmitter Development Section admitted that "sentiment in RCAM is somewhat divided [about FM], not as to the apparent technical advantages, but as to the advisability of its full adoption without further engineering, manufacturing and operating experience."[60] O. B. Hanson discovered that the entire RCA organization could provide no more than a couple of obsolete receivers. With some embarrassment, he therefore placed an order with two competitors, General Electric and Radio Engineering Laboratories, for seven FM sets, one of which was to be installed in the home of NBC's president, Lenox Lohr. In September the FCC approved NBC's application to construct a diminutive 1-kilowatt FM station in the Empire State Building.[61] Oddly, Hanson justified the expense, $12,000, as an opportunity to refute the claim that FM suppressed static. He pointed out at the time that "no real comparative tests have been made between frequency modulation and amplitude modulation on the same wave length."[62] This statement showed once more how Hanson and other RCA managers had completely lost touch with the still-evolving theory and normal practice of FM. In fact, FM pioneers had produced mountains of evidence, much of it published, that the Armstrong system suppressed static. Moreover, despite ample publicity about frequency modulation, Hanson mentioned none of the several other features that attracted broadcasters to FM, such as high fidelity and minimal interstation interference. At the moment of birth of the first commercial FM radio service in 1940, RCA had almost nothing to do with the delivery.

"Almost a Cakewalk": The FCC Creates Commercial FM

Within a year, FM radio would become a permanent fixture in American broadcasting, with RCA having almost nothing to say about the matter. On the morning of 18 March 1940, the FCC hearing to create a commercial FM broadcast service that John Shepard and FMBI had sought since October convened. Chairman James Lawrence Fly rapped his gavel to open what he declared as FM's "day in court," the largest assembly before the commission in its five-year history. Fly had to borrow a three-hundred-seat auditorium from the Interstate Commerce Commission, and one hundred people still had to listen to the proceedings from loudspeakers mounted outside.

To those who attended the two-week-long hearing, the greatest surprise was a lack of rancor. Twenty-nine organizations had asked to send representatives to testify, and reporters who had been covering wideband FM predicted that the hearing would continue Armstrong's "fight" or "battle" with RCA on behalf of his invention.[63] At the start of the hearing, the *New York Times* described the radio industry as "sharply divided" over FM, and *Broadcasting* noted that a "substantial portion" of the four hundred audience members "viewed FM as a prospective Frankenstein that might turn on their established station operations and introduce new competition of a character that might prove ruinous."[64] But after these skeptics heard FMBI witnesses forecast a transition period of "roughly 10 years," "this viewpoint appeared to subside."[65] Moreover, signs soon appeared that the commission would reach a favorable conclusion for the FM pioneers. When Armstrong, the first witness, played recordings inscribed on motion-picture film of AM and FM reception in Haddonfield that RCA had made in 1935, *Electronics* stated that "the advantage in favor of f-m was so marked, and the static so prominent on the a-m portions of the film that Chairman Fly asked that the final record be turned off before its conclusion, granting the demonstration as conclusive." Armstrong commanded such a strong position that he could easily afford to be unusually conciliatory, admitting that FM's lower static levels were partly "due to the use of the higher frequencies, inasmuch as natural static decreases roughly in proportion to the increase in frequency." But he also insisted that FM excelled at discriminating against man-made noises, an assertion that no one challenged.[66]

Several observers expected far more wrangling during the second week, and *Electronics* confidently promised "stiff opposition on the part of RCA." But Frank W. Wozencraft, RCA's chief counsel, urged the FCC to give regular FM service the "green light," thus "taking FM proponents wholly by surprise," reported *Broadcasting*.[67] Wozencraft made only two requests. First, he asked the commission to give a television channel other than No. 1 to FM, "since that channel is now in regular use." He also repeated RCA's suggestion that FM channels narrower than 200 kilocycles per second might suffice for practical purposes, although he undercut his position by conceding that a wider channel had a better signal-to-noise ratio.[68] Both of these ideas attracted almost no support. "If there is any real opposition to FM as a new commercial service to supplement rather than supplant the present standard broadcast structure employing amplitude modulation," Sol Taishoff, the editor of *Broadcasting*, declared, "it was not evident during the proceedings." And if any debate occurred, the FCC almost always ruled in favor of FMBI. "What was expected to be a battle royal between opponents and proponents of wide-band FM," Taishoff observed, "turned out to be almost a cakewalk for the disciples of Maj. Edwin H. Armstrong."[69]

In fact, FM's "trial" more closely resembled a two-week colloquium, as the commissioners interviewed one expert witness after another about technical specifications. This was understandable, given that the FCC would be the first agency to regulate the new medium. The panel's sharpest questions challenged the necessity of a 15,000-cycle audio bandwidth. Chief Engineer Jett asked Armstrong whether 10,000 or 11,000 might suffice, to which Armstrong responded that 15,000 cycles gives the greatest "naturalness" to reception. Commissioner T. A. M. Craven, Jett's predecessor, as chief engineer, asked the inventor much the same thing, perhaps seeking a way to narrow the Armstrong system's 200-kilocycle channel width. Armstrong essentially answered that reducing the audio bandwidth would have almost no effect on the channel width, an explanation that ended Craven's questions. At no time did any commissioner contradict the technical judgment of Armstrong or any other advocate of his system.[70]

Of course, the FCC's subsequent decisions were anticlimactic. On 20 May 1940 the panel established the commercial service that FMBI wanted on a band of spectrum from 42 to 50 megacycles, enough for forty channels. The lowest five, from 42 to 43 megacycles, were reserved for "educational stations on a regular broadcast basis," establishing a precedent for the 4 megacycles of noncommercial broadcasting on today's FM band.[71] The 200-kilocycle channel was retained, and television Channel 1, which had occupied 44 to 50 megacycles, was eliminated. The panel adopted none of RCA's proposals. During the past nearly seven decades, broadcast FM radio has continued to evolve; the FCC shifted the FM band to its present location in 1945, for instance, and in 1961 the commission authorized a method to broadcast stereophonic sound. But the FCC has never failed to preserve the essential standards of the technology, which Armstrong and other FM pioneers worked out during the late 1930s.

Conclusion

> By a process of evolution, [frequency modulation] may well supersede most of our existing system of radio before ten years or less.
>
> *American broadcast engineer, 1940*

> What's wrong with American FM?
>
> Popular Electronics, *1962*

This book has situated the history of FM between two complementary questions: Was frequency-modulation radio socially constructed? Or was it determined by natural law? The answer to both questions is yes, but the social origins of the technology exerted far more influence than did nature. Nature constrained what was technologically possible, ruling out narrowband FM, for example, by refusing to cooperate with those who counted on that method to solve the problem of spectrum congestion. But closing off one path still left a virtual infinity of other paths from which to choose. In other words, the "black-boxed" version of FM that we hear today was not the only possible result, or even the most likely or optimal outcome. In 1940 the FCC mandated a set of specifications to which FM broadcasters and transmitter and receiver manufacturers had to conform—namely, a 200-kilocycle channel width, a 150-kilocycle frequency swing, and a 15-kilocycle audio bandwidth. FM radio in America and many other countries is still based on these specifications. But no *technical* reason has ever existed to prevent the commission from diverging from that standard, choosing, for example a 100-kilocycle channel, a 50-kilocycle swing, and a 10-kilocycle audio bandwidth. The FM we hear today essentially matches the system that Armstrong and other FM pioneers had standardized by 1939 only because the FCC accepted their judgment in 1940.

The years of continuously shifting social factors, ranging from the organizational to the personal, strongly shaped the technology of FM radio. The amateur radio community played a key role in the long-term development of FM systems by educating a generation of boys about radio before World War I. Many adolescent hams grew up to become professional engineers, and nearly every professional engineer who did significant work with frequency modulation during the 1930s and 1940s entered the radio technology community as a youngster. Further, amateur radio clubs found and created forums for hams to hone their skills as public practitioners and engineers. The Radio Club of America sent Armstrong to the Hoover Radio Conferences that first met in 1922, and several members assisted him in exhibiting and promoting wideband FM in the 1930s. The amateur radio community also provided an alternative complimentary culture to the corporations that employed radio engineers. Radio clubs encouraged the free exchange of information, as opposed to companies, which placed a higher priority on the control of proprietary secrets. Thanks to those clubs, radio engineering was a relatively communal profession.

A number of changes in the context of radio during the twenties and thirties also shaped the present system of broadcast FM radio and helped determine when it appeared. None of these changes can be considered simply technological or social in the traditional overly narrow meanings of those words. The congestion of AM radio broadcasting triggered a conceptual move to the spectrum paradigm, which in turn shaped the debate about how radio in general should be regulated. Congestion and the new paradigm accounted as well for why engineers took another look at FM radiotelephony during the early 1920s, after twenty years of that technology's dormancy. Narrowband FM, for a brief time, signified for some people the best hope for curing congestion, but they soon discovered the futility of that idea. Nevertheless, for two reasons this disappointment marked one of the most valuable lessons in the history of FM: first, it taught that narrowband could never work; and, second, it inspired minded engineers at Westinghouse and RCAC to develop further the mathematical theory behind frequency modulation.

An even more significant accelerant to FM research originated entirely within the commercial context of radio. During the first half of the 1920s, engineers at General Electric, Westinghouse, and RCAC independently investigated frequency modulation, albeit while envisioning different purposes for the method. Although noncompetition contracts tied these companies together, the firms did not cooperate on FM research until they began to implement David Sarnoff's short-lived unification plan in 1928. By compelling GE and Westinghouse to share the results of their radio research with RCAC, unification effectively funneled almost everything that was known about FM to the latter organization, an intellectual windfall that spared RCAC engineers a great deal of spadework. Without unification, some sort of practical FM would have emerged from Westinghouse or RCAC (or perhaps another source), although any hypothetical system would have arrived perhaps decades after 1933, the year that Armstrong was issued his wideband patents. In retrospect, the union of Westinghouse and RCA research benefited Armstrong more than anyone.

If Armstrong was a great engineer, it was not because he was a genius, unless one considers his remarkably dogged ability to frame and reframe sometimes unworkable theories. For all the errors he committed while developing wideband FM—his theoretical misconception of balanced amplifiers, for instance, and his assumption that frequency-modulation radio would have no effect on static noise—he ranks as a superb designer of radio hardware. The singularity of his technological achievements existed, though, within a social, economic, and cultural context in which the theory and practice of frequency-modulation radiotelephony was already familiar to, if not mastered by, scores of his colleagues. Traditionally depicted as an independent inventor who single-handedly invented wideband FM, Armstrong in fact stood on the shoulders of earlier researchers who had labored, often successfully, on many problems associated with FM. This is not to minimize his several crucial improvements to the art, such as a radically wider frequency swing and a balanced-amplifier detector that replaced the old slope detector. But his invention of wideband FM and his development of a broadcast service depended on both indirect and direct assistance, especially from RCA. All during the 1920s and early 1930s, he exploited his unique access, as a consultant, shareholder, and friend, to the work and material support of RCAC engineers, who themselves had recently acquired valuable knowledge from Westinghouse and GE. Moreover, Armstrong rarely reciprocated, typically keeping his work secret until his patents were securely in hand. And when it came time to sell FM to the FCC, it was FM pioneers who did so, another instance of how Armstrong relied on communities of practitioners to further his goals and continue to shape frequency-modulation technology.

This study also overturns the fifty-year-old conventional explanation for Armstrong's greatest disappointment during the 1930s: his failure to obtain RCA's backing for wideband FM. Why RCA squandered its chance to acquire the rights to wideband FM is attributable to a number of causes, but the company's desire to protect its investments in AM technology was not one of them, as amplitude-modulation radio broadcasting accounted for only a tiny proportion of RCA's overall capitalization and income. Rather, RCA's hitherto puzzling behavior occurred chiefly because FM was at first so theoretically abstruse. Because no one, including Armstrong, foresaw until the spring of 1934 what advantages (such as static noise suppression and high fidelity) wideband FM would ultimately offer, the tests he and RCA carried out that year were not designed to confirm, let alone quantify, those advantages. Instead, RCA evaluated FM only with respect to the primary claim of Armstrong's patents—namely, that the system extended the service range of short-wave communications. The testing strategy, which narrowly focused on only this feature, created a distorted picture of Armstrong's invention, and no credible consensus developed within RCA about what he had actually invented. It was an acute, though understandable, blunder for which Armstrong himself shoulders much of the blame, because he misunderstood wideband FM radio, too. Still, the company might have sponsored the Armstrong system had a half dozen or so of its managers caught the error in time. Instead, RCA's engineer-managers continued obliviously both to play down the technical potential of wideband FM and, later, to underestimate the persuasive power of the ever-growing community of FM pioneers. As a result, RCA let FM get away.

Despite Armstrong's inability to win over RCA, this book elevates him as an exceptional "heterogeneous engineer"—that is, an individual who played the "social" side of the technology he promoted as skillfully as he played the side that is traditionally seen as technical.[1] One can understand this role and Armstrong's relationship with both RCA and the FM pioneers in terms of "actor-network theory," as conceived by sociologists of technology approximately fifteen years ago. From the early 1920s until 1936 Armstrong was associated with the "pre-existing network" of the RCA organization at large. When he failed to obtain RCA's backing, he created a new "social-technological" network by recruiting—sociologists would say "enrolling"—FM pioneers to help further improve the technology and to sell FM to the public. In doing so, Armstrong became a "dedicated network builder."[2] His most valuable recruit, John Shepard III, was also an experienced "network" builder in two senses of the term when Armstrong met him. Shepard had, of course, created the first (temporary) radio network as an experiment in 1923, and afterward, even while leading the FM movement, he presided over an organization of regional AM broadcasters that he had cofounded. By late 1939 Shepard headed FM Broadcasters, Inc., the group he had almost single-handedly created for the purpose of lobbying the FCC on behalf of the Armstrong system. Thanks to a strategy largely of Shepard's design, the commission gave FM an "enthusiastic green light" only a few months later.[3]

The history of FM radio helps drive another stake into the heart of technological determinism, which in 1985 Wajcman and Mackenzie asserted was "the single most influential theory of the relationship between technology and society."[4] The belief that the evolution of technology operates according to its own internal logic, or that external factors minimally bear on that evolution, has waned in popularity during the past twenty years, especially among scholars who study the history and sociology of science and technology. But technological determinism is nowhere near its deserved death, due partly to writers who still neglect interrelated technological and social contexts in modern life. According to Thomas Misa, someone who commits the error of technological determinism often adopts the wrong "perspective" in how he or she approaches the object of study. "Those historians (and others) adopting a 'macro' perspective are the ones who allow technology a causal role in historical change. They deploy the Machine to make history." As a partial solution, obviously, one can adopt—as does this book—a sufficiently "micro" perspective in which the "causal role for the Machine is not present and is not possible."[5] But while assuming a micro perspective one should not also revert to old-fashioned internalism, which disconnects technology from society in other ways. A historian must at least take into account the exogenous factors that continually shape the hardware that emerges from the experiences of technological practitioners. And without denying that technology "impacts" society, we must keep in mind that society *always* acts on technology as well. Indeed, this book examines the origins of a specific technology during a period when that technology influenced society minimally, if at all.

Finally, this book contributes to a long-standing debate in the field of science, technology, and society studies about the role of nature in technological innovation. As with technological determinism, the argument that nature plays no part in the construction of technology is made far less vigorously today than a few years ago, but the idea survives in many corners of academia. The history of FM provides ample empirical evidence that nature imposes limitations on what technology can and cannot do. Walter Vincenti has pointed out that when Thomas Edison was developing his electric lighting system from 1877 to 1882, at least two simple "non-negotiable" technical constraints imposed by the "real world" constrained Edison and his staff: Ohm's law, for current, voltage, and resistance ($I = V \div R$); and Joule's law, for electrical power, resistance, and current ($P = R \times I^2$). Vincenti avoids making a simple essentialist argument by allowing that these formulas are "human artefacts, subject to modification or coercion over time." But he also asserts that "in the absence of anything demonstrably better, power engineers have to take them—in fact, they think of them—as tantamount to the real world itself."[6]

Of course, *thinking* of a statement as "tantamount to the real world" does not ensure that the statement *is* tantamount to—or even descriptive of—the real world. But the fact that no one has documented a violation of Ohm's and Joule's laws—despite powerful social, economic, and technological incentives for doing so—constitutes compelling evidence that we sometimes just cannot interpret our way around a natural law, whether that law is socially constructed or not. Indeed, the history of FM provides a far stronger and more interesting role for nature than does Ohm's law: not only did FM inventors believe that they were constrained by the same *known* laws that governed Edison, but practical FM radio turned out to be constrained as well by *unknown* laws that researchers had to feel out for themselves. In three notable instances, those laws contradicted the previous expectations of researchers. First, narrowband FM advocates hoped that they could cure congestion on the AM broadcast band. Second, Armstrong believed, for a several years, that balanced amplifiers would reduce static noise. And, third, he was so certain that FM could not suppress static and that wider channels always brought greater noise levels that he memorialized his conviction in one of his famous wideband FM patents of 1933. In all three cases, testing demonstrated the original conceptions as wrong. No engineer, not even an Armstrong, Crosby, or Hansell, could bend frequency-modulation technology according to his ideology or other social values in contravention of the rules of the natural world.

Nearly seven decades have passed since the Federal Communications Commission established a commercial FM broadcast service, during which FM radio has continued to reflect the political, economic, and aesthetic values of society.[7] Today the FCC reserves a 4-megacycle portion of the FM broadcast band for nonprofit organizations, an artifact of Armstrong's strategy to bring as many AM broadcasters into the FM camp as he could afford. During the 1930s, he assessed commercial broadcasters relatively modest fees for using his fifteen FM patents—$5,000 for a 50,000-watt transmitter, for example. But educators received an essentially free ride because Armstrong charged them only one dollar.[8] As more and more colleges and universities applied for FM station licenses during the 1940s, New Deal reformers, including most FCC commissioners at the time, increasingly saw the Armstrong system as "radio's second chance" for education, since the FRC had withdrawn the great bulk of AM licenses for educational broadcasters in the 1920s and 1930s.[9] Over the long term, Armstrong's strategy worked. Today hundreds of educational and other nonprofit organizations in America operate FM stations.

Wideband FM was the first high-fidelity mass medium, and the FCC has traditionally fostered further improvements in its quality. In 1961 the commission authorized FM stations to use a multiplexing technology in their transmitters that made possible stereophonic radio broadcasts.[10] Although multiplexing slightly degraded the signal-to-noise ratio, on balance most listeners accepted the trade-offs. At first the improvement appealed mainly to "hi-fi" buffs, many of whom played classical music exclusively to show off FM's wider audio frequency response and greater dynamic (loudness) range.

FM broadcasting spread erratically during the fifties. Ironically, it only puttered along in America, attracting more listeners in urban than in rural areas, but never really challenging its older rival until nearly forty years after Armstrong was issued his patents. Not until 1983 did American FM stations outnumber AM stations.[11] FM often found far more popularity abroad, especially in Europe and the USSR. As a Soviet broadcast engineer pointed out during the early 1960s, "You Americans had the technical ability to produce FM, but it takes us and the Europeans to show you how to use it." In 1961 only 912 American FM stations were licensed, a *decrease* of 10 percent from 1950, and a fraction of the number of AM stations. By contrast, European stations during the same period climbed from 4 in number to approximately 1,000. One reason was that European radio tended to be government operated, and therefore listeners had fewer alternative sources for programs on AM bands. But also, European listeners heard more variety on their FM receivers than did Americans. During the 1950s one of the most popular networks in Germany—FM or AM—was the American Armed Forces Radio Service, established during the post–World War II occupation, and which aired jazz, popular, and even rock-and-roll music.[12]

FM in America, however, became associated narrowly with a highbrow culture that some characterized as overly devoted to "educational radio" and classical music. "If the AM band has become the home of rock-and-roll," one American critic complained in 1962, "much of the FM band is nothing more than a classical juke box."[13] Woody Allen made a similar point fifteen years later in his 1977 film, *Annie Hall.* While attempting to impress his new girlfriend, the insecure protagonist, Alvy, realizes that he has resorted to pretentious jargon about modern photography. "Christ," he tells himself, "I sound like FM radio!"[14]

Making fun of FM's highbrow reputation worked in 1977, but not ten years later, for within that decade FM radio would undergo another transformation. As historians of postwar radio have explained, several historical events removed the taint of high culture from American FM. Most important, rock music of the 1960s and 1970s, much of which demanded audio fidelity beyond the technological limits of AM, found a home on the underpopulated FM band.[15] Eventually, many FM stations became commercially successful, but other problems now afflict the FM band. As the ownership of American media becomes increasingly concentrated and less diverse, listeners are served up more and more predictable and ever blander programming. National broadcasters have invested in technology that enables the production of generic programs in, say, Los Angeles, that are disguised as locally produced shows in dozens of distant cities.

No reason exists to assume that the present sorry state of the medium is permanent, but can we know FM's future? There is some truth in Howard Armstrong's declaration that "the best way to look into the future is to look at the past."[16] But the history of FM indicates that forecasting the future of technology can be, at the very least, tricky. In October 1940 the chief engineer of radio station WOR, John R. Poppele, predicted before a meeting of the Radio Club of America that, "stemming from such a rosy present, it seems inevitable that FM will have an illustrious career of steady growth. . . . By a process of evolution, it may well supersede most of our existing system of radio before ten years or less."[17] Later events soon proved Poppele dead wrong, and now, at the dawn of the twenty-first century, developments in communications technology becloud FM's future more than ever. Digital modulation could render both FM *and* AM broadcasting obsolete within a decade or two. Or perhaps the FCC's present policy of encouraging low-power FM will revive community broadcasting by fostering thousands of short-range, low-wattage transmitters.[18] Indeed, the difficulty of predicting the future of FM demonstrates the principal argument of this study—namely, that a technology is no more inevitable than the historical events that continually shape it.

APPENDIX

FM-Related Patents, 1902–1953

Patents are listed in order of filing date. Except for Poulsen's Danish Patent No. 5,590, all patents shown are U.S. patents, which can be viewed on the U.S. Patent and Trademark Office Web page, www.uspto.gov/.

Inventor	Title	Notes	Assignee	Patent No.	Application Date	Issue Date
Ehret, Cornelius D.	Art of Transmitting Intelligence	First FM patents. Antifading.	None	785,803	10 Feb. 1902	28 Mar. 1905
Ehret, Cornelius D.	System of Transmitting Intelligence	First FM patents. Antifading.	None	785,804	10 Feb.1902	28 Mar. 1905
Poulsen, Vlademar	Method for Generating Alternating Current with High Frequencies	Arc oscillator using FSK	None	5,590 (Danish)	9 Sept. 1902	3 Apr. 1903
Hammond, John Hays, Jr.	Radio Telegraphy and Telephony	FSK for telegraphy. Amplitude modulation for telephony.	None	1,320,685	29 May 1912	4 Nov. 1919
Hewitt, Peter Cooper	System of Electrical Distribution	FM transmitter with conducting gas or vapor oscillator	None	780,999	13 Dec. 1913	25 Feb. 1919
Nelson, Edward L.	Modulating and Transmitting System	FSK with audion amplifier	WE	1,349,729	8 Mar. 1918	17 Aug. 1920
Day, Albert V. T.	Method and Means for Electrical Signaling and Control		None	1,885,009	25 Jan. 1919	25 Oct. 1932
Taylor, Albert Hoyt	Simultaneous Transmission or Reception of Speech and Signals	FSK telegraphy, with AM of two wavelengths	None	1,376,051	10 Apr. 1919	26 Apr. 1921
Nyman, Alexander	Combined Wireless Sending and Receiving System	Uses a Poulsen arc	WEM	1,615,645	15 July 1920	25 Jan. 1927
Little, Donald G.	Wireless Telephone System		WEM	1,595,794	30 June 1921	10 Aug. 1926
Conrad, Frank	Wireless Telephone System	FM transmitter	WEM	1,528,047	15 Mar. 1922	3 Mar. 1925

Inventor	Title	Notes	Assignee	Patent No.	Application Date	Issue Date
Purington, Ellison S.	Radiant Signaling System	Transmitter and receiver. "Wobble from a wave length of 505 to 526 meters, which would cause the mean value of the wave length to be 515.5 meters." Corresponds to a center frequency of 581,851 cps, with a frequency swing of 23,703 cps.	JHH	1,599,586	27 Apr. 1922	14 Sept. 1926
Mertz, Pierre	Electrical Transmission of Pictures	Transmitter and receiver. FM for wired transmission of pictures. Limiter.	AT&T	1,548,895	26 Jan. 1923	11 Aug. 1925
Sivian, Leon J.	Means and Method for Signaling by Electric Waves	Transmitter and receiver. Describes a narrowband FM system with a channel width of 4,000 cycles.	WE	1,847,142	5 Dec. 1923	1 Mar. 1932
Schelleng, John C.	Electric Wave Signaling System	Transmitter and receiver	WE	1,653,878	22 Dec. 1923	27 Dec. 1927
Shanck, Roy B.	Picture Transmitting System	Transmitter and receiver	AT&T	2,115,917	12 Mar. 1925	3 May 1938
Hartley, Ralph V. L.	Electric-Wave-Modulating System		WE	1,633,016	7 July 1925	21 June 1927
Wright, George Maurice, and Smith, Sidney Bertram	Radio Transmission and Reception of Pictures	Transmitter and receiver. Limiting in FM for pictures. Antifading. Antistatic.	RCA	1,964,375	20 Feb. 1926	26 June 1934
Long, Maurice B.	Electrical Transmission System	Transmitter and receiver	WE	1,977,683	22 May 1926	23 Oct. 1934
Mohr, Franklin	Transmission System	Transmitter only	WE	1,715,561	26 May 1926	4 June 1929
Coleman, John B.	Transmitting System	Transmitter only. Electronic coupled telegraph keying.	WEM	1,920,296	7 Aug. 1926	1 Aug. 1933
Alexanderson, Ernst F. W.	Transmission of Pictures	Transmitter and receiver. Antifading.	GE	1,830,586	9 Aug. 1926	3 Nov. 1931

Inventor	Title	Notes	Assignee	Patent No.	Application Date	Issue Date
Demarest, Charles S.	Signaling System	Transmitter and receiver. Limiting in FM receiver.	AT&T	2,047,312	1 Dec. 1926	14 July 1936
Conrad, Frank	Radio Communication System	Transmitter and receiver. FM balanced discriminator.	WEM	2,057,640	17 Mar. 1927	13 Oct. 1936
Chireix, Henri	Means for Radio Communication	Transmitter and receiver. Efficient phase modulation.	None	1,882,119	6 May 1927	11 Oct. 1932
Armstrong, Edwin H.	Radio Telephone Signaling	Transmitter and receiver. Narrowband FM.	None	1,941,447	18 May 1927	26 Dec. 1933
Trouant, Virgil E.	Radio Transmitting System	Electronic modulation. Improvement on Conrad Patent No. 2,057,640.	WEM	1,953,140	18 June 1927	3 Apr. 1934
Peterson, Harold O.	Signaling by Frequency Modulation	Transmitter and receiver. Use of FM to narrow channel width. Antifading.	RCA	1,789,371	12 July 1927	20 Jan. 1931
Armstrong, Edwin H.	Radio Signaling System	FSK	None	2,082,935	6 Aug. 1927	8 June 1937
Hansell, Clarence W.	Communication by Frequency Variation	Transmitter and receiver. Wobbling. Multiplexing. Balanced frequency modulator. Spectrum conservation. Receiver limiting. Magnetic wobbler or capacity variation or resistance modulation.	RCA	1,819,508	11 Aug. 1927	18 Aug. 1931
Usselman, George Lindley	Frequency Modulation	Transmitter and receiver. Uses balanced circuits in transmitter to suppress harmonics and ensure a constant amplitude.	RCA	1,794,932	1 Sept. 1927	3 Mar. 1931
Trouant, Virgil E.	Radiotransmitting System	Transmitter only. Crystal-controlled oscillator. Narrowband FM.	WEM	1,872,364	8 Oct. 1927	16 Aug. 1932
Albersheim, Walter S.	Method and Means for Signaling by Frequency Fluctuation	Transmitter and receiver. Improved linearity when modulating with small frequency deviation.	RCA	1,999,176	28 Jan. 1928	30 Apr. 1935

Inventor	Title	Notes	Assignee	Patent No.	Application Date	Issue Date
Jones, Lester L.	Variable Relay Condenser	Nonmechanical high-speed variable condenser for FM	None	1,777,410	6 Mar. 1928	7 Oct. 1930
Hansell, Clarence W.	Oscillation Generation	A more stable FM transmitter. Antifading.	RCA	1,787,979	23 Mar. 1928	6 Jan. 1931
Hansell, Clarence W.	Frequency Modulation	Transmitter only. Narrower channel width and antifading. Wobbler. Greater efficiency.	RCA	1,830,166	23 Mar. 1928	3 Nov. 1931
Trouant, Virgil E.	Radio Station	Transmitter and receiver	WEM	1,861,462	3 May 1928	7 June 1932
Potter, Ralph K.	Electrooptical Image-Producing System	Transmitter and receiver. Phase modulation for pictures.	AT&T	1,777,016	19 May 1928	30 Sept. 1930
Hansell, Clarence W.	Signaling	Transmitter and receiver. Multiplex signaling on a high frequency carrier. Limiter in receiver.	RCA	2,103,847	2 Oct. 1928	28 Dec. 1937
Hansell, Clarence W.	Signaling	Transmitter and receiver. Antifading. Limiter in receiver. Heterodyne in transmitter. Spectrum conservation.	RCA	1,803,504	5 Oct. 1928	5 May 1931
Beverage, Harold H.	Signaling	Transmitter and receiver. Spectrum conservation. Antifading.	RCA	1,849,608	19 Nov. 1928	15 Mar. 1932
Hansell, Clarence W.	Detection of Frequency Modulated Signals	Receiver only. Circuit to "limit the amplitude of the received alternating energy so greatly that the output current is practically square in wave form."	RCA	1,813,922	30 Jan. 1929	14 July 1931
Hansell, Clarence W.	Detection of Frequency Modulated Signals		RCA	1,867,567	1 Feb. 1929	19 July 1932
Hansell, Clarence W.	Detection of Frequency Modulated Signals	Receiver only. Limiter circuit. Alternative to slope detector. Makes more linear the resonance curve used in detection.	RCA	1,938,657	1 Feb. 1929	12 Dec. 1933

Inventor	Title	Notes	Assignee	Patent No.	Application Date	Issue Date
Ranger, Richard Howland	Wobbled Frequency Superheterodyne System	Transmitter and receiver	RCA	1,830,242	22 Mar. 1929	3 Nov. 1931
Hammond, John Hays, Jr.	Transmission of Light Sequences by Frequency Variation	Transmitter and receiver. FM, for both radio and light waves.	None	1,977,438	18 May 1929	16 Oct. 1934
Hammond, John Hays, Jr.	Transmission of Intelligence by Frequency Variation	Transmitter and receiver	None	1,977,439	18 May 1929	16 Oct. 1934
Hammond, John Hays, Jr.	Transmission of Light Variations by Frequency Variations	Transmitter only. FM of radio and light waves.	None	2,036,869	18 May 1929	7 Apr. 1936
Hansell, Clarence W.	Frequency Changer	Used to wobble a transmitter frequency	RCA	1,874,982	20 June 1929	30 Aug. 1932
Böhm, Otto	Signaling by Frequency Modulation	Transmitter only. Simple FM modulation of a crystal oscillator.	TGFDT	1,874,869	8 July 1929	30 Aug. 1932
Dome, Robert B.	Frequency Modulation	Transmitter only. Gives as a "practical example" a frequency swing of 20,000 cycles, after multiplying by 100.	GE	1,917,102	22 July 1929	4 July 1933
Schriever, Otto	Signaling	Transmitter and receiver	TGFDT	1,911,091	3 Sept. 1929	23 May 1933
Van Der Pol, Balthasar	Device for Modulating the Frequency of Electric Oscillations	Transmitter only. Uses light-sensitive cells to frequency-modulate high-frequency oscillations.	RCA	1,876,109	20 Nov. 1929	6 Sept. 1932
Hansell, Clarence W.	Signaling	Transmitter only. Limits the channel width of the transmitter with a bandpass filter.	RCA	1,849,620	16 Jan. 1930	15 Mar. 1932
Day, Albert V. T.	Carrier Wave Signaling	Transmitter and receiver	None	2,164,032	14 Feb. 1930	27 June 1939

Inventor	Title	Notes	Assignee	Patent No.	Application Date	Issue Date
Hansell, Clarence W.	Detection of Frequency Modulated Signals	Receiver only	RCA	1,922,290	14 May 1930	15 Aug. 1933
Hansell, Clarence W.	Frequency Modulation	Transmitter only. Long-line frequency controlled oscillator.	RCA	2,027,975	25 June 1930	14 Jan. 1936
Roberts, Walter van B.	Frequency Modulation	Transmitter only	RCA	1,917,394	10 July 1930	11 July 1933
Armstrong, Edwin H.	Radio Signaling System	Transmitter and receiver	None	1,941,066	30 July 1930	26 Dec. 1933
Goldstine, Hallan Eugene	Modulation of Oscillations	Increased efficiency	RCA	2,067,081	31 Jan. 1931	5 Jan. 1937
Wasserman, Marian George	Frequency Multiplication	Applicable to FM and phase modulation	CGTSF	1,964,373	18 Feb. 1931	26 June 1934
Lindenblad, Nils E.	Modulation	Transmitter only. Eliminates AM components from FM components and vice versa. Electronic modulation of "ultra-short-wave" oscillations.	RCA	1,938,749	27 Mar. 1931	12 Dec. 1933
Crosby, Murray G.	Reception of Phase Modulated Waves	Receiver only	RCA	2,114,335	25 Sept. 1931	19 Apr. 1938
Wolcott, Carl Frederick	Communication System		None	1,972,964	28 Sept. 1931	11 Sept. 1934
Crosby, Murray G.	Phase Modulation	Transmitter and receiver	RCA	2,081,577	23 Jan. 1932	25 May 1937
Crosby, Murray G.	Phase Modulation Receiver	Limits "the frequency band of the receiver to the band occupied by the signal. This results in minimum noise and interference."	RCA	2,101,703	23 Jan. 1932	7 Dec. 1937
Crosby, Murray G.	Frequency or Phase Modulation	Transmitter only	RCA	2,085,739	30 Apr. 1932	6 July 1937

Inventor	Title	Notes	Assignee	Patent No.	Application Date	Issue Date
Roosenstein, Hans Otto	Modulation	Transmitter only. High-speed modulation.	TGFDT	2,001,891	5 May 1932	21 May 1935
Hansell, Clarence W.	Signaling	Transmitter and receiver. Antifading.	RCA	1,999,902	13 May 1932	30 Apr. 1935
Usselman, George Lindley	Phase and Frequency Modulation	Transmitter only. Antifading.	RCA	2,036,165	8 June 1932	31 Mar.1936
Crosby, Murray G.	Signal Receiver	FM and phase modulation	RCA	2,229,640	20 June 1932	28 Jan. 1941
Crosby, Murray G.	Signal Receiver	FM and phase modulation	RCA	2,230,212	20 June 1932	28 Jan. 1941
Crosby, Murray G.	Frequency Modulation	Transmitter only. Push-pull oscillator.	RCA	2,032,403	9 July 1932	3 Mar. 1937
Hansell, Clarence W.	Electrical Circuits	Transmitter only. Elimination of amplitude modulation in a phase modulation or FM system.	RCA	1,999,190	31 Oct. 1932	30 Apr. 1935
Conklin, James W.	Frequency Modulation	Transmitter only. Magnetron oscillator, followed by amplitude limiter.	RCA	1,965,332	17 Dec. 1932	3 July 1934
Turner, Alfred H.	Secret Signaling	Transmitter only	RCA	2,026,758	21 Dec. 1932	7 Jan. 1936
Lindenblad, Nils E.	Short Wave Signaling	Transmitter only	RCA	2,052,888	14 Jan. 1933	1 Sept. 1936
Conklin, James W.	Receiver	Phase modulation and FM	RCA	2,091,271	17 Jan. 1933	31 Aug. 1937
Armstrong, Edwin H.	Radiosignaling	Wideband FM	None	1,941,068	24 Jan. 1933	26 Dec. 1933
Armstrong, Edwin H.	Radiosignaling	Wideband FM	None	1,941,069	24 Jan. 1933	26 Dec. 1933
George, Ralph W.	Receiving Means	Carrier frequencies of 70 cm (428 mc). Uses IF amplifier 600 kilocycles wide, "of sufficient width to make its use in television desirable."	RCA	2,035,745	27 Apr. 1933	31 Mar. 1936

Inventor	Title	Notes	Assignee	Patent No.	Application Date	Issue Date
Philpott, La Verne R.	Facsimile Transmission System	Transmitter only	WEM	2,070,312	21 June 1933	9 Feb. 1937
Hansell, Clarence W.	Variable Reactance Modulator Circuit	Rapid modulation of FM and phase modulation systems	RCA	2,121,737	24 July 1933	21 June 1938
Usselman, George Lindley	Frequency Modulation Apparatus	Transmitter only	RCA	2,030,125	25 Sept. 1933	11 Feb. 1936
Bechmann, Rudolf, and Elstermann, Herbert	Frequency Modulation	Transmitter only	TGFDT	2,076,289	10 Nov. 1933	6 Apr. 1937
Lindenblad, Nils E.	Modulation	Transmitter only. "Used in the transcontinental test between Bolinas and Riverhead."	RCA	2,054,431	29 Nov. 1933	15 Sept. 1936
Chaffee, Joseph G.	Modulating System for Ultra Short Waves	Transmitter only. Adaptable for FM.	BTL	2,038,992	15 Dec. 1933	28 Apr. 1936
Crosby, Murray G.	Receiving System	Phase modulation and FM	RCA	2,060,611	23 Dec. 1933	10 Nov. 1936
Crosby, Murray G.	Receiver	Receiver only. FM, AM, or phase modulation.	RCA	2,064,106	27 Dec. 1933	15 Dec. 1936
Crosby, Murray G.	Frequency Modulation Phase Modulation Receiver	Receiver only	RCA	2,112,881	28 Dec. 1933	5 Apr. 1938
Conklin, James W.	Frequency Modulation Detection		RCA	2,095,314	23 Mar. 1934	12 Oct. 1937

Inventor	Title	Notes	Assignee	Patent No.	Application Date	Issue Date
Crosby, Murray G.	Phase and Amplitude Modulation Receiver		RCA	2,019,446	9 Apr. 1934	29 Oct. 1935
Crosby, Murray G.	Phase and Frequency Modulation Wave Receiving System		RCA	2,087,429	6 June 1935	20 July 1937
Armstrong, Edwin H.	Radio Transmitting System		None	2,063,074	14 Sept. 1935	8 Dec. 1936
Armstrong, Edwin H.	Radio Transmitting System		None	2,098,698	14 Sept. 1935	9 Nov. 1937
Armstrong, Edwin H.	Radio Signaling System		None	2,104,011	14 Sept. 1935	4 Jan. 1938
Armstrong, Edwin H.	Multiplex Radio Signaling System		None	2,104,012	14 Sept. 1935	4 Jan. 1938
Armstrong, Edwin H.	Radio Receiving System		None	2,116,501	14 Sept. 1935	10 May 1938
Conrad, Frank	Receiving System	Simultaneous AM and FM reception	WEM	2,151,747	14 Sept. 1935	28 Mar. 1939
Usselman, George Lindley	Oscillation Generator and Frequency Modulator		RCA	2,160,466	5 Oct. 1935	30 May 1939
Crosby, Murray G.	Detection of Frequency Modulated Signals		RCA	2,071,113	17 Oct. 1935	16 Feb. 1937
Seeley, Stuart William	Frequency Variation Response Circuits	Applicable to FM detection	RCA	2,121,103	17 Oct. 1935	21 June 1938
Roberts, Walter van B.	Electric Phase Controlling Circuit		RCA	2,215,127	9 Nov. 1935	26 July 1938
Chaffee, Joseph G.	Receiving System for Frequency Modulated Waves		BTL	2,118,161	24 Dec. 1935	24 May 1938

Inventor	Title	Notes	Assignee	Patent No.	Application Date	Issue Date
Chaffee, Joseph G.	Reception of Frequency Modulated Waves		BTL	2,075,503	26 Mar. 1936	30 Mar. 1937
Armstrong, Edwin H.	Radio Receiving System		None	2,116,502	25 Aug. 1936	10 May 1938
Hansell, Clarence W.	Frequency Modulation Circuits	Improved balanced receiver	RCA	2,179,182	27 Nov. 1936	7 Nov. 1939
Bown, Ralph	Frequency Modulation		BTL	2,212,338	28 Apr. 1938	20 Aug. 1940
Chaffee, Joseph G.	Radio Repeater	Improved repeaters for FM	BTL	2,148,532	28 Apr. 1938	28 Feb. 1939
Roder, Hans	Frequency Modulation System	Allows a weaker signal to break into a stronger signal	GE	2,270,899	12 Nov. 1938	27 Jan. 1942
Armstrong, Edwin H.	Means and Method for Relaying Frequency Modulated Signals		None	2,264,608	12 Jan. 1940	2 Dec. 1941
Armstrong, Edwin H.	Radio Rebroadcasting System		None	2,276,008	12 Jan. 1940	10 Mar. 1942
Armstrong, Edwin H.	Frequency Modulation System		None	2,290,159	12 Jan. 1940	21 July 1942
Armstrong, Edwin H.	Means for Receiving Radio Signals		None	2,318,137	12 Jan. 1940	4 May 1943
Armstrong, Edwin H.	Frequency-Modulated Carrier Signal Receiver		None	2,540,643	12 Jan. 1940	6 Feb. 1951
Armstrong, Edwin H.	Current Limiting Device		None	2,295,323	2 Aug. 1940	8 Sept. 1942
Armstrong, Edwin H.	Method and Means for Transmitting Frequency Modulated Signals		None	2,315,308	2 Aug. 1940	30 Mar. 1943

Inventor	Title	Notes	Assignee	Patent No.	Application Date	Issue Date
Armstrong, Edwin H.	Means and Method for Relaying Frequency Modulated Signals		None	2,275,486	25 Sept. 1940	10 Mar. 1942
Roberts, Walter van B.	Frequency Modulation		RCA	2,289,041	10 Oct. 1940	7 July 1942
Armstrong, Edwin H.	Frequency Modulation Signaling System		None	2,323,698	12 Oct. 1940	6 July 1943
Armstrong, Edwin H.	Radio Signaling		None	2,602,885	30 Mar.1946	8 July 1952
Armstrong, Edwin H.	Frequency Modulation Multiplex System		None	2,630,497	1 June 1949	3 Mar. 1953

Introduction • *What Do We Know about FM Radio?*

Epigraph. Congress, Senate, Committee on Interstate Commerce, *A Bill to Amend the Communications Act of 1934, and for Other Purposes*, 78th Cong., 1st sess., 3 Nov.–16 Dec. 1943, 679.

1. Ken Burns, Geoffrey C. Ward, and Tom Lewis, *Empire of the Air: The Men Who Made Radio*, film directed by Ken Burns (Los Angeles, Calif.: Pacific Arts Video, 1992); Lawrence P. Lessing, *Man of High Fidelity: Edwin Howard Armstrong* (Philadelphia: J. B. Lippincott, 1956); *Man of High Fidelity: Edwin Howard Armstrong*, 2d ed., with new foreword and final chapter (New York: Bantam Books, 1969). Unless otherwise noted, all page citations are from the 1956 edition.

2. See Donald MacKenzie and Judy Wajcman, eds., *The Social Shaping of Technology*, 2d ed. (Buckingham: Open University Press, 1999); Wiebe E. Bijker, Thomas Parke Hughes, and Trevor J. Pinch, eds., *The Social Construction of Technological Systems: New Directions in the Sociology and History of Technology* (Cambridge, Mass.: MIT Press, 1987); Angela Lakwete, *Inventing the Cotton Gin: Machine and Myth in Antebellum America* (Baltimore: Johns Hopkins University Press, 2003); David F. Noble, *America by Design: Science, Technology, and the Rise of Corporate Capitalism* (New York: Oxford University Press, 1977); Merritt Roe Smith and Leo Marx, eds., *Does Technology Drive History? The Dilemma of Technological Determinism* (Cambridge, Mass.: MIT Press, 1994); Nathan Rosenberg, *Inside the Black Box: Technology and Economics* (Cambridge: Cambridge University Press, 1982). The argument for incremental technological change is at least several decades old. See S. Colum Gilfillan, *The Sociology of Invention* (Chicago: Folliett, 1935; repr., Cambridge, Mass.: MIT Press, 1970); George Basalla, *The Evolution of Technology* (New York: Cambridge University Press, 1988).

3. This study combines the categories of phase modulation and frequency modulation. The differences between the two are trivial for the purposes of this study.

4. In 1960 the Conférence générale des poids et measures defined the Hertz as the new unit of measurement for frequency, but here I use cycles per second (cps) to conform to the great majority of sources used in this study.

5. It should be noted that, strictly speaking, the concept of "instantaneous frequency" is mathematically and practically problematic when the frequency is changing. I use the concept only to sketch out the basics of frequency-modulation theory for the layperson.

6. Lessing, *Man of High Fidelity*, 293.

7. Ibid., 308.

8. Ibid., 258.

9. Ibid., 226, 309.

10. Ibid., 309, 234, 240, 296, 299.

11. Ibid., 11. The foreword to the 1969 edition of *Man of High Fidelity*, ix, makes a stronger claim, offering a historical, legal, and personal "vindication" of FM radio.

12. Hugh Slotten disputes the traditional claim that in 1945 the FCC transferred the commercial broadcast FM service from 42–50 megacycles to 88–108 megacycles out of a motivation to "cripple" FM. Hugh Richard Slotten, "'Rainbow in the Sky': FM Radio, Technical Superiority, and Regulatory Decision-Making, 1936–1948," in *Radio and Television Regulation: Broadcast Technology in the United States, 1920–1960* (Baltimore: Johns Hopkins University Press, 2000), 113–44.

13. The other book that focuses on the history of FM radio is Don V. Erickson, *Armstrong's Fight for FM Broadcasting: One Man vs. Big Business and Bureaucracy* (University: University of Alabama Press, 1973). See also Charles A. Siepmann, *Radio's Second Chance* (Boston: Atlantic-Little, Brown Books, 1946); W. Rupert Maclaurin, *Invention and Innovation in the Radio Industry* (New York: Macmillan, 1949), 183–90, 228–31; Christopher H. Sterling, "Second Service: Some Keys to the Development of FM Broadcasting," *Journal of Broadcasting* 15 (Spring 1971): 181–94; Lawrence Lessig, *Free Culture: How Big Media Uses Technology and Law to Lock Down Culture and Control Creativity* (New York: Penguin, 2004), 3–6. Almost all Web pages that recount the history of FM radio cite Lessing and often Erikson. See, for example, Jorge Amador, "Edwin Armstrong: Genius of FM Radio" (Apr. 1990), www.thefreemanonline.org/columns/edwin-armstrong-genius-of-fm-radio/, Web page of the Foundation for Economic Education, viewed 1 Apr. 2009. One Web-based history of FM radio seems to embellish Lessing, but cites no sources. "Edwin Howard Armstrong: The Creator of FM Radio," Web page of the First Electronic Church of America, http://fecha.org/armstrong.htm, viewed 1 Apr. 2009.

14. Thomas Parke Hughes, *American Genesis: A Century of Invention and Technological Enthusiasm* (New York: Viking Press, 1989); Tom Lewis, *Empire of the Air: The Men Who Made Radio* (New York: Edward Burlingame Books, 1991). Lewis's book was made in collaboration with Ken Burns. See Burns, Ward, and Lewis, *Empire of the Air*; Susan J. Douglas, "The FM Revolution," in *Listening In: Radio and the American Imagination, from Amos 'n' Andy and Edward R. Murrow to Wolfman Jack and Howard Stern* (New York: Times Books, 1999), 256–83. Douglas suggests the possibility of an alternative to "Armstrong's conspiracy theory" (262, n. 14).

15. Christopher H. Sterling and Michael C. Keith, *Sounds of Change: A History of FM Broadcasting in America* (Chapel Hill: University of North Carolina Press, 2008).

16. John M. Staudenmaier, "Recent Trends in the History of Technology," *American Historical Review* 95 (June 1990): 715–25; and *Technology's Storytellers: Reweaving the Human Fabric* (Cambridge, Mass.: MIT Press, 1985).

17. Smith and Marx, introduction to *Does Technology Drive History?*, ix–xv.

18. Thomas Parke Hughes, "The Seamless Web: Technology, Science, Etcetera, Etcetera," *Social Studies of Science* 16 (May 1986): 281–92.

19. Donald MacKenzie and Judy Wajcman, "Introductory Essay: The Social Shaping of Technology," in MacKenzie and Wajcman, *The Social Shaping of Technology*, 3–27.

20. Nathan Rosenberg, "Technological Interdependence in the American Economy," *Technology and Culture* 20 (Jan. 1979): 25; reprinted in Rosenberg, *Inside the Black Box*, 55.

21. Susan J. Douglas, "Popular Culture and Populist Technology: The Amateur Operators, 1906–1912," in *Inventing American Broadcasting, 1899–1922* (Baltimore: Johns Hopkins University Press, 1987), 187–215.

Chapter One • AM and FM Radio before 1920

Epigraph. Valdemar Poulsen, "A Method of Producing Undamped Electric Oscillations and Its Employment in Wireless Telegraphy," *Electrician* 58 (16 Nov. 1906): 166–68. The author thanks Hans Buhl for providing copies of several articles related to the Poulsen arc, and for pointing out that Poulsen opposed the very method he invented. Emails to author 18–20 July 2001. For an English-language summary of Dr. Buhl's Ph.D. dissertation, see "The Arc Transmitter—A Comparative Study of the Invention, Development and Innovation of the Poulsen System in Denmark, England and the United States," Web page of the Steno Museum for the History of Science and Medicine, University of Aarhus, www.stenomuseet.dk/person/hb.ukref.htm, viewed 1 Apr. 2009.

1. Hugh G. J. Aitken, *Syntony and Spark: The Origins of Radio* (Princeton: Princeton University Press, 1976); Sungook Hong, *Wireless: From Marconi's Black Box to the Audion* (Cambridge, Mass.: MIT Press, 2001), 19.

2. Edward W. Constant II, *The Origins of the Turbojet Revolution* (Baltimore: Johns Hopkins University Press, 1980); Thomas Parke Hughes, *Networks of Power: Electrification in Western Society, 1850–1930* (Baltimore: Johns Hopkins University Press, 1983), esp. 15–16.

3. Reginald Aubrey Fessenden, "Wireless Telephony," *Proceedings of the Institute of Electrical Engineers* 27 (29 June 1908): 1283–1358.

4. Hong, *Wireless*, 96 (emphasis in original).

5. Cornelius D. Ehret, "Art of Transmitting Intelligence," U.S. Patent No. 785,803, application date: 10 Feb. 1902, issue date: 28 Mar. 1905; and, "System of Transmitting Intelligence," U.S. Patent No. 785,804, application date: 10 Feb. 1902, issue date: 28 Mar. 1905.

6. Valdemar Poulsen, "Fremgangsmaade til Frembringelse af Vekselstromme med højt Svingningstal" (Method for Generating Alternating Current with High Frequencies), Danish Patent No. 5,590, application date: 9 Sept. 1902, issue date: 3 Apr. 1903.

7. Raymond F. Guy, "F-M and U-H-F," *Communications* 28 (Aug. 1943): 30; Ehret is also mentioned in a footnote in Christopher H. Sterling, "Second Service: Some Keys to the Development of FM Broadcasting," *Journal of Broadcasting* 15 (Spring 1971): 181–94. Sterling and Keith also discuss Ehret in their 2008 book. Christopher H. Sterling and Michael C. Keith, *Sounds of Change: A History of FM Broadcasting in America* (Chapel Hill: University of North Carolina Press, 2008).

8. *Armstrong v. Emerson Radio and Phonograph Corporation*, 179 F. Supp. 95, Southern District of New York, 1959.

9. Eugene S. Ferguson, "The Tools of Visualization," in *Engineering and the Mind's Eye* (Cambridge, Mass.: MIT Press, 1992), 75–113.

10. Ehret, U.S. Patent Nos. 785,803, 785,804.

11. See, for example, Frederick Emmon Terman, *Radio Engineering* (New York: McGraw-Hill, 1937), 587.

12. For an early example of a limiter circuit, see Pierre Mertz, "Electrical Transmission of Pictures," U.S. Patent No. 1,548,895, application date: 26 Jan.1923, issue date: 11 Aug. 1925. This patent describes a wired frequency-modulation picture transmission system.

13. Poulsen, Danish Patent No. 5,590. In June 1903 Poulsen applied for an American patent for the same invention. Valdemar Poulsen, "Method of Producing Alternating Currents with a High Number of Vibrations," U.S. Patent No. 789,449, application date: 19 June 1903, issue date: 9 May 1905.

14. Aitken shows three photographs of three such alternators, all rated at 200 kilowatts output. Hugh G. J. Aitken, *Continuous Wave: Technology and American Radio, 1900–1932* (Princeton: Princeton University Press, 1985), 313, 324, and 423.

15. Hong, *Wireless,* 164.

16. Poulsen, "A Method of Producing Undamped Electric Oscillations."

17. Ibid., 167.

18. "Poulsen Wireless Station at Lyngby," *Modern Electrics* 1 (June 1908): 81. This article states that "good operators can work with less than 1% difference of the wave length."

19. Poulsen, "A Method of Producing Undamped Electric Oscillations," 167.

20. Cyril F. Elwell, "The Poulsen System of Wireless Telephony and Telegraphy," *Journal of Electricity, Power and Gas* 24 (2 Apr. 1910): 293–97, quotation on p. 294.

21. Lee de Forest, "Recent Developments in the Work of the Federal Telegraph Company," *PIRE* 1 (Mar. 1913): 37–57, quotation on p. 39.

22. Elihu Thomson, "Method and Means for Producing Alternating Currents," U.S. Patent No. 500,630, application date: 18 July 1892, issue date: 4 July 1893.

23. John Grant, "Experiments and Results in Wireless Telephony," *American Telephone Journal* 15 (26 Jan. 1907): 49; Fessenden, "Wireless Telephony."

24. Elmer E. Bucher, *Practical Wireless Telegraphy: A Complete Text Book for Students of Radio Communication,* rev. ed. (New York: Wireless Press, 1921), 267.

25. Reginald Aubrey Fessenden, "Apparatus for Signaling by Electromagnetic Waves," U.S. Patent No. 706,747, application date: 28 Sept. 1901, issue date: 12 Aug. 1902.

26. Gary L. Frost, "The Two Careers of Reginald Aubrey Fessenden" (Master's thesis, University of North Carolina–Chapel Hill, 1993).

27. Alexander Nyman, "Combined Wireless Sending and Receiving System," U.S. Patent No. 1,615,645, application date: 15 July 1920, issue date: 25 Jan. 1927. Technically, Albert Day beat Nyman by more than a year, filing for his phase-modulation invention in early 1919. But the thirteen years separating Day's filing and issue dates suggest that the final document differed considerably from the original application. Albert V. T. Day, "Method and Means for Electrical Signaling and Control," U.S. Patent No. 1,885,009, application date: 25 Jan. 1919, issue date: 25 Oct. 1932.

28. Murray G. Crosby, "Notes on FM Lecture by Major Armstrong given in the A.I.E.E. sponsored course in frequency modulation on Oct. 14, 1940," 16 Oct. 1940, box 183, AP.

29. The number of female amateur radio operators has never been determined. I have discovered no women practitioners who worked directly with FM radio technology before World War II.

30. Greenleaf Whittier Pickard, "Means for Receiving Intelligence by Communicated by Electric Waves," U.S. Patent No. 836,531, application date: 30 Aug. 1906, issue date: 20 Nov. 1906; Henry C. Dunwoody, "Wireless-Telegraph System," U.S. Patent No. 837,616, application date: 23 Mar. 1906, issue date: 4 Dec. 1906.

31. Susan J. Douglas, "Popular Culture and Populist Technology: The Amateur Operators, 1906–1912," in *Inventing American Broadcasting, 1899–1922* (Baltimore: Johns Hopkins University Press, 1987), 188–215.

32. Ibid. For an examination of masculinity and amateur radio in post–World War I America, see Kristin Haring, *Ham Radio's Technical Culture* (Cambridge, Mass.: MIT Press, 2006).

33. "The Wireless Club," *Electrician and Mechanic* 19 (Sept. 1908): 137.

34. Hugo Gernsback, "Wireless and the Amateur: A Retrospect," *Modern Electrics* 4 (Feb. 1913): 1143.

35. Robert A. Morton, "The Amateur Wireless Operator," *Outlook* 94 (15 Jan. 1910): 131.

36. "The Radio League of America: A Retrospect," *Electrical Experimenter* 3 (Dec. 1915): 381.

37. Harold H. Beverage and Harold O. Peterson, Electrical Engineers, an oral history conducted in 1968 and 1973 by Norval Dwyer, IEEE History Center, Rutgers University, New Brunswick, N.J.

38. Ibid. "DX" referred to the practice of listening to distant radio stations.

39. Radio Club of America, "A History of the Radio Club of America," Radio Club of America Web page, www.radio-club-of-america.org/history.php, viewed 1 Apr. 2009, adapted from George E. Burghard, *The Twenty-Fifth Anniversary Year Book* (New York: Radio Club of America, 1934).

40. W. E. D. Stokes Jr. remained an amateur and a member of the Radio Club of America. In 1989 he suffered a stroke, and he died on 10 March 1992, when he was ninety-six years old. Houston Stokes, telephone interview by Gary L. Frost, 27 Feb. 2002.

41. "The First Amateur Radio Club in America," *Radio Broadcast* 2 (Jan. 1923): 222.

42. Ibid. See "Godley to England to Copy Transatlantics," *QST* 5 (Oct. 1921): 29; George E. Burghard, "Eighteen Years of Amateur Radio," *Radio Broadcast* 2 (Aug. 1923): 290.

43. Gernsback, "Wireless and the Amateur," 1143.

44. Ibid.

45. "Boy Amateur, Assailing Wireless Bill, Predicts an 'Air Trust,'" *New York Herald,* 29 Apr. 1910.

46. Radio Club of America, "History of the Radio Club of America."

47. "First Amateur Radio Club in America," 222.

48. Burghard, "A History of the Radio Club of America."

49. Paul F. Godley, "Prof. Armstrong's System—What It Means: Frequency Modulation Plan Furnishes Plenty of Food for Thought on Future of the Broadcast Industry," *BBA* 11 (1 July 1936): 72. Burghard was the club's president in 1923. See "Radio Device Ending Fading, Static Reported: Armstrong Perfects Method Increasing Range of Ultra-Short Wave Broadcasting," *New York Herald Tribune,* 26 Apr. 1935.

50. Armstrong thanked Styles, Shaughnessy, Burghard, and Runyon for their help in his seminal paper on FM radio. Edwin Howard Armstrong, "A Method of Reducing Disturbances in Radio Signaling by a System of Frequency Modulation," *PIRE* 24 (May 1936): 689–740.

Chapter Two • Congestion and Frequency-Modulation Research, 1913–1933

Epigraphs. Lose M. Ezzy [pseud.], "I'm Forever Losing Signals," *QST* 4 (Sept. 1920): 43. H. P. Davis to David Sarnoff, 11 Mar. 1931, box 275, microfilm reel, "RCA-1 C.S. & M: #1 to #1419," frame 595, AP.

1. Erik Barnouw, *A Tower in Babel: A History of Broadcasting in the United States to 1933* (New York: Oxford University Press, 1966; repr. 1978). For a detailed history of congestion, see Marvin R. Bensman, *The Beginning of Broadcast Regulation in the Twentieth Century* (Jefferson, N.C.: McFarland, 2000).

2. Gleason L. Archer, *History of Radio to 1926* (New York: American Historical Society, 1938), 393–97.

3. Hugh G. J. Aitken, "Allocating the Spectrum: The Origin of Radio Regulation," *Technology and Culture* 35 (Oct. 1994): 686–716; Bensman, *Broadcast Regulation,* 84.

4. Bensman, *Broadcast Regulation,* 84.

5. Benjamin Gross, "How to Demonstrate F-M: Details of Public Demonstration at R. H. White Department Store, with Complete Radio Script of Program Transmitted by W1XOJ," *FM* 1 (Nov. 1940): 16.

6. *Radio Act of 1912,* Pub. L. 62-264, 13 Aug. 1912, sec. 4.

7. Sungook Hong, "Hertzian Optics and Wireless Telegraphy," in *Wireless: From Marconi's Black Box to the Audion* (Cambridge, Mass.: MIT Press, 2001), 1–24.

8. "Wireless Transmission of News," *Telephony* 71 (30 Dec. 1916): 32.

9. Edward Bellamy, *Looking Backward, 2000–1887* (Boston: Houghton Mifflin, 1887), 112–15; "Telephonic News Distribution," *Electrical World* 21 (18 Mar. 1893): 212; "Telephone Newspaper," *Electrical World* 22 (4 Nov. 1893): 362.

10. Arthur F. Colton, "The Telephone Newspaper—New Experiment in America," *Telephony* 62 (30 Mar. 1912): 391–92.

11. Bureau of Navigation, Department of Commerce, *Radio Service Bulletin* 61 (1 May 1922): 23–30.

12. K. B. Warner, "The Washington Radio Conference," *QST* 5 (Apr. 1922): 7.

13. *New York Times,* 10 Feb. 1922, quoted in Bensman, *Broadcast Regulation,* 48.

14. K. B. Warner, "The Second National Radio Conference," *QST* 6 (May 1923): 12.

15. Bureau of Navigation, Department of Commerce, *Radio Service Bulletin* 72 (2 Apr. 1923): 11 (emphasis added).

16. Bensman, *Broadcast Regulation*, 136. The Bureau of Navigation's new preference for frequencies was reported in the *Radio Service Bulletin* 72 (2 Apr. 1923): 11–12.

17. According to Bensman, *Broadcast Regulation*, 82, "it was some 15 years later that references to meters in various [government] publications finally stopped."

18. Federal Radio Commission, *Annual Report of the Federal Radio Commission to the Congress of the United States for the Fiscal Year Ended June 30, 1927* (Washington, D.C.: Government Printing Office, 1927), 6.

19. For histories of the commercialization of American radio during the 1920s and 1930s, see Susan Smulyan, *Selling Radio: The Commercialization of American Broadcasting, 1920–1934* (Washington, D.C.: Smithsonian Institution Press, 1994); Robert W. McChesney, *Telecommunications, Mass Media, and Democracy: The Battle for the Control of U.S. Broadcasting, 1928–1935* (New York: Oxford University Press, 1993).

20. Edgar H. Felix, "Will New Transmitting Methods Be the Remedy?" *Radio Broadcast* 13 (May 1928): 5.

21. IEEE History Center, "George W. Pierce," IEEE History Center Web page, www.ieeeghn.org/wiki/index.php/George_W._Pierce, viewed 27 Mar. 2009.

22. James E. Brittain, "Scanning the Past: John R. Carson and Conservation of Radio Spectrum," *Proceedings of the Institute of Electrical and Electronic Engineers* 84 (June 1996): 909. SSB found wide usage in many kinds of telecommunications, and continues to do so today, but it proved impractical for broadcasting. Building a stable single-sideband receiver cost too much for the home radio market during the 1920s; and, besides, the huge number of double-sideband receivers made impractical a changeover on a national scale.

23. Leon J. Sivian, "Means and Method for Signaling by Electric Waves." U.S. Patent No. 1,847,142, application date: 5 Dec. 1923, issue date: 1 Mar. 1932, assigned to Western Electric Company.

24. Clarence W. Hansell, "Phase and Frequency Modulation Applied to Short Wave Communications," 6 Jan. 1932, box 275, microfilm reel, "RCA-7 C.S. & M: 191-M-5923 to 226-M-7212," frame 6329, AP.

25. John Renshaw Carson, "Notes on the Theory of Modulation," *PIRE* 10 (Feb. 1922): 57–64, quotations on pp. 57, 59 (emphasis in original). The *Oxford English Dictionary*, 2d ed., s.v., "frequency modulation," correctly cites this article as the earliest usage of "frequency modulation" and "amplitude modulation."

26. Carson mentioned no specific frequencies in his article. For the purpose of explanation I have chosen to use 100 cps as a modulating frequency and a 1,000,000 cps carrier.

27. Carson, "Notes on the Theory of Modulation."

28. Lawrence P. Lessing, *Man of High Fidelity: Edwin Howard Armstrong* (Philadelphia: J. B. Lippincott, 1956), 199.

29. For examples of similar misinterpretations of Carson, see W. Rupert Maclaurin, *Invention and Innovation in the Radio Industry* (New York: Macmillan, 1949), 185;

and Lessing, *Man of High Fidelity*, 199. Clarence Hansell consistently distorted Carson's argument during the 1930s. In 1932 he stated that "Carson . . . even went so far as to state: 'Consequently this type of modulation (frequency modulation) inherently distorts without any compensating advantages whatsoever.'" Clearly, Hansell implied that Carson was referring to all kinds of frequency modulation. Hansell, "Phase and Frequency Modulation Applied to Short Wave Communications," 6 Jan. 1932. Seven years later, Hansell mentioned "the AT&T attempt to condemn frequency modulation telephony for all time." Hansell to C. H. Taylor, 5 Jan. 1939, box 275, microfilm reel, "RCA-4 C.S. & M: #3435 to #4696," frame 3583, AP. Howard Armstrong grew increasingly contemptuous of mathematical theories of FM, partly because he also believed Carson had erred. In 1943 Armstrong declared in a criticism of author August Hund, that "Carson's [1922] paper . . . made definite predictions of great disadvantages for FM. Whether [Hund], who is essentially a mathematician, resents the fact that in the case of the FM System the experimental method gave the mathematicians a terrible showing up, or whether he has some other motive for covering up the truth, I do not know." This statement indicates that Armstrong either did not understand what Carson had actually said or deliberately misinterpreted Carson. Armstrong to Paul A. de Mars, 4 Apr. 1943, box 7, AP.

30. Lessing, *Man of High Fidelity*, 199–200; Edwin Howard Armstrong, "Mathematical Theory vs. Physical Concept," *FM and Television* 4 (Aug. 1944): 11; John R. Ragazzini, "Creativity in Radio: Contributions of Major Edwin Howard Armstrong," *Journal of Engineering Education* 45 (Oct. 1954): 112–19, reprinted in John W. Morrisey, ed., *The Legacies of Edwin Howard Armstrong* (n.p.: Radio Club of America, 1990), 41; Don V. Erickson, *Armstrong's Fight for FM Broadcasting: One Man vs. Big Business and Bureaucracy* (University: University of Alabama Press, 1973).

31. "John R. Carson," *PIRE* 16 (July 1928): 862; Brittain, "Scanning the Past: John R. Carson and Conservation of Radio Spectrum."

32. Hansell, "Phase and Frequency Modulation Applied to Short Wave Communications," 6 Jan. 1932.

33. Lessing, *Man of High Fidelity*, 200.

34. Albert Hoyt Taylor, "Simultaneous Transmission or Reception of Speech and Signals," U.S. Patent No. 1,376,051, application date: 10 Apr. 1919, issue date: 26 Apr. 1921.

35. John Hays Hammond Jr., "Radio Telegraphy and Telephony," U.S. Patent No. 1,320,685, application date: 29 May 1912, issue date: 4 Nov. 1919.

36. Peter Cooper Hewitt, "System of Electrical Distribution," U.S. Patent No. 1,295,499, application date: 13 Dec. 1913, issue date: 25 Feb. 1919.

37. Albert V. T. Day, "Method and Means for Electrical Signaling and Control," U.S. Patent No. 1,885,009, application date: 25 Jan. 1919, issue date: 25 Oct. 1932.

38. John Hays Hammond Jr., "Transmission of Light by Frequency Variation," U.S. Patent No. 1,977,438, application date: 18 May 1929, issue date: 16 Oct. 1934; and "Transmission of Light Variations by Frequency Variations," U.S. Patent No. 2,036,869, application date: 18 May 1929, issue date: 7 Apr. 1936.

39. Ernst F. W. Alexanderson, "Transmission of Pictures," U.S. Patent No. 1,830,586,

application date: 9 Aug. 1926, issue date: 3 Nov. 1931, assigned to General Electric; Otto Böhm, "Signaling by Frequency Modulation," U.S. Patent No. 1,874,869, application date: 8 July 1929, issue date: 30 Aug. 1932, assigned to Telefunken Gesellschaft für Drahtlose Telegraphie mbH; Robert B. Dome, "Frequency Modulation," U.S. Patent No. 1,917,102, application date: 22 July 1929, issue date: 4 July 1933, assigned to General Electric Company; Otto Schriever, "Signaling," U.S. Patent No. 1,911,091, application date: 3 Sept. 1929, issue date: 23 May 1933, assigned to Telefunken Gesellschaft für Drahtlose Telegraphie mbH; Marian George Wasserman, "Frequency Multiplication," U.S. Patent No. 1,964,373, application date: 18 Feb. 1931, issue date: 26 June 1934, assigned to Compagnie Générale de Télégraphie Sans Fil; Hans Otto Roosenstein, "Modulation," U.S. Patent No. 2,001,891, application date: 5 May 1932, issue date: 21 May 1935, assigned to Telefunken Gesellschaft für Drahtlose Telegraphie m; Rudolf Bechmann and Herbert Elstermann, "Frequency Modulation," U.S. Patent No. 2,076,289, application date: 10 Nov. 1933, issue date: 6 Apr. 1937, assigned to Telefunken Gesellschaft für Drahtlose Telegraphie mbH.

40. The facsimile patents assigned to AT&T before 1933 comprise the following: Pierre Mertz, "Electrical Transmission of Pictures," U.S. Patent No. 1,548,895, application date: 26 Jan. 1923, issue date: 11 Aug. 1925; Ralph K. Potter, "Electrooptical Image-Producing System," U.S. Patent No. 1,777,016, application date: 19 May 1928, issue date: 30 Sept. 1930; Roy B. Shanck, "Picture Transmitting System," U.S. Patent No. 2,115,917, application date: 12 Mar. 1925, issue date: 3 May 1938. Facsimile patents assigned to Western Electric are: Franklin Mohr, "Transmission System," U.S. Patent No. 1,715,561, application date: 26 May 1926, issue date: 4 June 1929; Maurice B. Long, "Electrical Transmission System," U.S. Patent No. 1,977,683, application date: 22 May 1926, issue date: 23 Oct. 1934.

41. Charles S. Demarest, "Signaling System," U.S. Patent No. 2,047,312, application date: 1 Dec. 1926, issue date: 14 July 1936.

42. For a short discussion of Westinghouse's experimental broadcasting work, see Maclaurin, *Invention and Innovation in the Radio Industry*, 174.

43. H. P. Davis, "The Early History of Broadcasting in the United States," in Graduate School of Business Administration, Harvard University, *The Radio Industry: The Story of Its Development* (Chicago: A. W. Shaw, 1928): 216; Maclaurin, *Invention and Innovation in the Radio Industry*, 174.

44. George W. Pierce, "Electrical System," U.S. Patent No. 2,133,642, application date: 25 Feb. 1924, issue date: 18 Oct. 1938.

45. Alexander Nyman, "Combined Wireless Sending and Receiving System," U.S. Patent No. 1,615,645, application date: 15 July 1920, issue date: 25 Jan. 1927; Donald G. Little, "Wireless Telephone System," U.S. Patent No. 1,595,794, application date: 30 June 1921, issue date: 10 Aug. 1926, assigned to Westinghouse. Little's patent appears to contain a typographical error: "The object of my invention," it reads, "is to provide a modulating system of the above indicated character, wherein the radiated energy is modulated in amplitude rather than in frequency." This sentence implies that the invention is an amplitude-modulation system, but the remainder of the document clearly describes frequency modulation.

46. FM-related patents filed during the 1920s and assigned to Westinghouse include (in order of their filing dates): Nyman, U.S. Patent No. 1,615,645; Little, U.S. Patent No. 1,595,794; Frank Conrad, "Wireless Telephone System," U.S. Patent No. 1,528,047, application date: 15 Mar. 1922, issue date: 3 Mar. 1925; John B. Coleman, "Transmitting System," U.S. Patent No. 1,920,296, application date: 7 Aug. 1926, issue date: 1 Aug. 1933; Frank Conrad, "Radio Communication System," U.S. Patent No. 2,057,640, application date: 17 Mar. 1927, issue date: 13 Oct. 1936; Virgil E. Trouant, "Radio Transmitting System," U.S. Patent No. 1,953,140, application date: 18 June 1927, issue date: 3 Apr. 1934; Trouant, "Radiotransmitting System," U.S. Patent No. 1,872,364, application date: 8 Oct. 1927, issue date: 16 Aug. 1932; Trouant, "Radio Station," U.S. Patent No. 1,861,462, application date: 3 May 1928, issue date: 7 June 1932; La Verne R. Philpott, "Facsimile Transmission System," U.S. Patent No. 2,070,312, application date: 21 June 1933, issue date: 9 Feb. 1937.

47. C. W. Horn to Ralph R. Beal, 20 Oct. 1939, box 275, microfilm reel, "RCA-1 C.S. & M: #1 to #1419," frame 391, AP.

48. Ibid.

49. Coleman, U.S. Patent No. 1,920,296; Conrad, U.S. Patent No. 2,057,640; Trouant, U.S. Patent No. 1,953,140; Trouant, U.S. Patent No. 1,872,364; Trouant, U.S. Patent No. 1,861,462.

50. Trouant, U.S. Patent No. 1,872,364.

51. Horn to Beal, 20 Oct. 1939.

52. "Half-Million Will Be Spent for Tests: Shepard Starts 50 kw. Plant for Armstrong Experiments," *BBA* 14 (15 Jan. 1938): 15.

53. Horn to Beal, 20 Oct. 1939.

54. Davis, "Early History," 219.

55. Mary Texanna Loomis, *Radio Theory and Operating for the Radio Student and Practical Operator*, 4th, rev. ed. (Washington, D.C.: Loomis, 1928), 490–91.

56. Edgar H Felix, "Will New Transmitting Methods Be the Remedy?" *Radio Broadcast* 13 (May 1928): 5.

57. John Harmon, "Frequency Modulation: A Possible Cure for the Present Congestion of the Ether," *Wireless World* 29 (22 Jan. 1930): 89.

58. David F. Noble, *America by Design: Science, Technology, and the Rise of Corporate Capitalism* (New York: Oxford University Press, 1977), 88.

59. Ibid., 92.

60. Ibid., 93.

61. Barnouw, *Tower in Babel*, 186.

62. Hansell, "Phase and Frequency Modulation Applied to Short Wave Communications," 6 Jan. 1932.

63. Harold O. Peterson, "Signaling by Frequency Modulation," U.S. Patent No. 1,789,371, application date: 12 July 1927, issue date: 20 Jan. 1931, assigned to RCA. This patent only barely qualifies as a radiotelephony invention, as it chiefly focuses on a method of FSK telegraphy.

64. Hansell, "Phase and Frequency Modulation Applied to Short Wave Communications," 6 Jan. 1932.

65. Clarence W. Hansell, "Communication by Frequency Variation," U.S. Patent No. 1,819,508, application date: 11 Aug. 1927, issue date: 18 Aug. 1931, assigned to RCA; "Oscillation Generation," U.S. Patent No. 1,787,979, application date: 23 Mar. 1928, issue date: 6 Jan. 1931, assigned to RCA; "Signaling," U.S. Patent No. 2,103,847, application date: 2 Oct. 1928, issue date: 28 Dec. 1937, assigned to RCA; and "Signaling," U.S. Patent No. 1,803,504, application date: 5 Oct. 1928, issue date: 5 May 1931, assigned to RCA.

66. "Frequency Modulation in Radio Broadcasting," *Proceedings of the Radio Club of America* 16 (July 1939): 2.

Chapter Three • RCA, Armstrong, and the Acceleration of FM Research, 1926–1933

Epigraph. Beverage to T. J. Boerner, 26 Feb. 1932, box 275, microfilm reel, "RCA-7 C.S. & M: 191-M-5923 to 226-M-7212," frame 6320, AP.

1. Lawrence P. Lessing, *Man of High Fidelity: Edwin Howard Armstrong* (Philadelphia: J. B. Lippincott, 1956), 219.

2. Alan S. Douglas, "Who Invented the Superheterodyne?" in John W. Morrisey, ed., *The Legacies of Edwin Howard Armstrong* (n.p.: Radio Club of America, 1990), 123–42.

3. Lessing, *Man of High Fidelity*, 98–99.

4. Ibid., 131. Armstrong's regeneration circuit was the first electronic device that generated continuous radio waves. The superheterodyne circuit greatly simplified the tuning of a radio receiver. The "superhet" also made possible the mass production of inexpensive home radio receivers.

5. Ibid., 146. Ironically, superregeneration, the invention that Armstrong sold for the most money, was never widely used. Lessing also states that Armstrong "chose a good peak of $114 a share at which to sell a block of his R.C.A. stock just before the Crash, and emerged with a big profit unscathed" (ibid., 182). For additional information about Armstrong's personal wealth, see Carl Dreher, "E. H. Armstrong: The Hero as Inventor," *Harper's*, April 1956, 58, reprinted in Morrisey, *The Legacies of Edwin Howard Armstrong*, 11.

6. "Harold Beverage, Electrical Engineer," an oral history conducted in 1992 by Frederick Nebecker, IEEE History Center, Rutgers University, New Brunswick, N.J.

7. Harry Tunick to Armstrong, 1 Sept. 1931, box 162, envelope, "Conklin Setters; also RCA etc. Tunick & Sadenwater," AP.

8. Tunick to Armstrong, 19 Oct. 1931, box 162, envelope, "Conklin Setters; also RCA etc. Tunick & Sadenwater," AP.

9. Edwin H. Armstrong, "Radio Telephone Signaling," U.S. Patent No. 1,941,447, application date: 18 May 1927, issue date: 26 Dec. 1933.

10. Harold O. Peterson, "Signaling by Frequency Modulation," U.S. Patent No. 1,789,371, application date: 12 July 1927, issue date: 20 Jan. 1931, assigned to RCA.

11. Hugh G. J. Aitken, *Continuous Wave: Technology and American Radio, 1900–1932* (Princeton: Princeton University Press, 1985), 450–52.

12. Eugene Lyons, *David Sarnoff: A Biography* (New York: Harper & Row, 1966), 155–61.

13. Unknown RCA author, "Minutes of Meeting for Development Coordination," 1 Aug. 1929, box 275, AP.

14. Shelby's assessment is quoted in P. D. McKeel, "Description of Westinghouse Frequency Modulation Receiver," Aug.–Sept. 1930, box 275, AP.

15. Crosby to Beverage, 18 Dec. 1930, box 275, microfilm reel, "RCA-7 C.S. & M: 191-M-5923 to 226-M-7212," frame 7127, AP.

16. V. D. Landon, "An Analysis of Frequency Modulation," ca. 7 Sept. 1929, box 275, microfilm reel, "RCA-7 C.S. & M: 191-M-5923 to 226-M-7212," frame 7164, AP.

17. Hansell to C. H. Taylor, 7 Sept. 1929, box 275, microfilm reel, "RCA-7 C.S. & M: 191-M-5923 to 226-M-7212," frame 7163, AP.

18. Murray G. Crosby, "Theoretical and Experimental Analysis of the Transmitted Wave Form of a Frequency Modulation System," memorandum, 14 June 1930, box 275, AP.

19. Beverage to Alfred Norton Goldsmith, 21 July 1930, box 275, AP; Goldsmith to Beverage, 6 Aug. 1930, box 275, AP.

20. Beal to Hansell, 21 Oct. 1930, box 275, microfilm reel, "Cravath, Swaine & Moore Folder #190M: Index 90 to #5922," frame 5271, AP.

21. Beverage to Beal, 13 Jan. 1931, box 275, microfilm reel, "Cravath, Swaine & Moore Folder #190M: Index 90 to #5922," frame 5089, AP; also, box 275, microfilm reel, "RCA-7 C.S. & M: 191-M-5923 to 226-M-7212," frame 7122, AP.

22. Hansell to Conklin, "RD-1105—Frequency Modulation Experiments," memorandum, 27 June 1931, box 159, AP. Armstrong's transmitter and receiver were patented in Edwin H. Armstrong, "Radio Signaling System," U.S. Patent No. 1,941,066, application date: 30 July 1930, issue date: 26 Dec. 1933.

23. Hansell to Conklin, 27 June 1931.

24. Ibid.

25. Ibid.

26. Ibid.

27. Ibid.

28. Ibid.

29. Beal to Beverage, 7 July 1931, box 275, AP; copy of telegram, "McKesson, Manila to Conklin, Bolinas," undated attachment to Hansell to Armstrong, 16 July 1931, box 161, AP.

30. Beal to Beverage, 7 July 1931.

31. Armstrong to Hansell, 22 July 1931, box 162, envelope, "Conklin Setters; also RCA etc. Tunick & Sadenwater," AP.

32. Crosby to Armstrong, 8 Mar. 1932, box 160, AP; microfilmed copy in box 275, microfilm reel, "Cravath, Swaine & Moore Folder #190M: Index 90 to #5922," frame 5425, AP.

33. Beverage to T. J. Boerner, 26 Feb. 1932, box 275, microfilm reel, "RCA-7 C.S. & M: 191-M-5923 to 226-M-7212," frame 6320, AP.

34. Hansell to Beverage, 6 Jan. 1932, box 275, microfilm reel, "RCA-7 C.S. & M: 191-M-5923 to 226-M-7212," frame 6326, AP.

35. Clarence W. Hansell, "Phase and Frequency Modulation Applied to Short Wave Communications," 6 Jan. 1932, box 275, microfilm reel, "RCA-7 C.S. & M: 191-M-5923 to 226-M-7212," frame 6329, AP.

36. Hansell to Beverage, 6 Jan. 1932.

37. Beverage to Tunick, 8 Apr. 1932, box 160, AP.

38. Murray G. Crosby, "Frequency Modulation Propagation Characteristics," *PIRE* 24 (June 1936): 898–913.

39. Armstrong's paper was originally presented to the IRE on 6 Nov. 1935, Crosby's two days earlier. Edwin Howard Armstrong, "A Method of Reducing Disturbances in Radio Signaling by a System of Frequency Modulation," *PIRE* 24 (May 1936): 689–740.

40. Armstrong repeated Hansell's distortion of the critique of frequency modulation that Carson published in 1922. In 1932 Hansell wrote that "Dr. [Frank] Conrad was credited with statements to the effect that frequency modulation did not produce side frequencies as in amplitude modulation and would therefore permit a great increase in the number of broadcasting stations. The incorrectness of this assumption was pointed out by Dr. John R. Carson, who even went so far as to state: 'Consequently this type of modulation (frequency modulation) inherently distorts without any compensating advantages whatsoever.'" Hansell, "Phase and Frequency Modulation Applied to Short Wave Communications," 6 Jan. 1932. Six years later, Armstrong wrote in his famous paper on FM, "The subject of frequency modulation seemed forever closed with Carson's final judgment, rendered after a thorough consideration of the matter, that 'Consequently this method of modulation inherently distorts without any compensating advantages whatsoever.'" Armstrong, "A Method of Reducing Disturbances," 690.

Chapter Four • The Serendipitous Discovery of Staticless Radio, 1915–1935

Epigraphs. Walpole to Horace Mann, 28 Jan. 1754, quoted in Theodore G. Remer, *Serendipity and the Three Princes, from the Peregrinaggio of 1557* (Norman: University of Oklahoma Press, 1965), 6 (emphasis in original manuscript). The author thanks Richard Boyle, whose articles in the *Sunday Times* (Colombo, Sri Lanka) called attention to Walpole's original sense of the word "serendipity," as well as to Remer's book, which makes the same point. See Richard Boyle, "Serendipity and the Three Princes, Part One," *Sunday Times,* 30 July 2000, and "Serendipity and the Three Princes, Part Two," *Sunday Times,* 6 Aug. 2000.

1. For a list of "serendipity patterns" in science and technology, see Pek van Andel, "Anatomy of the Unsought Finding: Serendipity: Origins, History, Domains, Traditions, Appearances, Patterns and Programmability," *British Journal for the Philosophy of Science* 45 (June 1994): 631–48. Andel reaffirms earlier scholars' observations that the role of chance in discovery has been recognized by scientists and scholars since

the ancient Greeks. The best known aphorism about the role of chance in scientific discovery is Louis Pasteur's "Chance favors the prepared mind," although van Andel points out that this is a slight mistranslation from what Pasteur actually said, namely: "In the sciences of observation, chance favors only prepared minds" (ibid., 634–635). Remer examines instances of "accidental sagacity" in scientific and technological discovery, such as John Morehead's discovery of an inexpensive method of producing acetylene. Theodore G. Remer, "The Nature of Serendipity," in *Serendipity and the Three Princes, from the Peregrinaggio of 1557* (Norman: University of Oklahoma Press, 1965), 167–77.

2. Lawrence P. Lessing, *Man of High Fidelity: Edwin Howard Armstrong* (Philadelphia: J. B. Lippincott, 1956), 71.

3. Ibid., 196.

4. The linear balanced amplifier was also called, depending on its specific configuration and usage, a differential amplifier, a push-pull amplifier, or a class AB amplifier.

5. John Renshaw Carson, "Method and Means for Signaling with High-Frequency Waves," U.S. Patent No. 1,449,382, application date: 1 Dec. 1915, issue date: 27 Mar. 1923, assigned to AT&T; and "Duplex Translating-Circuits," U.S. Patent No. 1,343,307, application date: 5 Sept. 1916, issue date: 15 June 1920, assigned to AT&T.

6. Clarence W. Hansell, "Coupling," U.S. Patent No. 1,751,996, date of application: 18 Jan. 1927, date of issue: 25 Mar. 1930, assigned to RCA; "Oscillation Generation," U.S. Patent No. 1,787,979, application date: 23 Mar. 1928, issue date: 6 Jan. 1931, assigned to RCA; "Signaling," U.S. Patent No. 1,803,504, application date: 5 Oct. 1928, issue date: 5 May 1931, assigned to RCA; "Communication by Frequency Variation," U.S. Patent No. 1,819,508, application date: 11 Aug. 1927, issue date: 18 Aug. 1931, assigned to RCA; "Frequency Modulation," U.S. Patent No. 1,830,166, application date: 23 Mar. 1928, issue date: 3 Nov. 1931, assigned to RCA; "Detection of Frequency Modulated Signals," U.S. Patent No. 1,867,567, application date: 1 Feb. 1929, issue date: 19 July 1932, assigned to RCA; "Frequency Multiplier and Amplifier," U.S. Patent No. 1,878,308, application date: 23 Mar. 1927, issue date: 20 Sept. 1932, assigned to RCA; "Detection of Frequency Modulated Signals," U.S. Patent No. 1,922,290, application date: 14 May 1930, issue date: 15 Aug. 1933, assigned to RCA; "Frequency Modulation," U.S. Patent No. 2,027,975, application date: 25 June 1930, issue date: 14 Jan. 1936, assigned to RCA; "Signaling," U.S. Patent No. 2,103,847, application date: 2 Oct. 1928, issue date: 28 Dec. 1937, assigned to RCA; and "Signaling," U.S. Patent No. 2,167,480, application date: 2 Nov. 1927, issue date: 25 July 1939, assigned to RCA.

7. Edwin Howard Armstrong, "Some Recent Developments in the Audion Receiver," *PIRE* 3 (Sept. 1915): 215–47, reprinted in John W. Morrisey, ed., *The Legacies of Edwin Howard Armstrong* (n.p.: Radio Club of America, 1990), 189–216; also reprinted in *Proceedings of the IEEE* 85 (Apr. 1997): 685–97.

8. Edwin H. Armstrong. "Operating Features of the Audion," *Electrical World* 64 (12 Dec. 1914): 1149–52. For the testy exchange between Armstrong and de Forest, see the discussion that follows Armstrong, "Some Recent Developments."

9. Armstrong, "Some Recent Developments," 234, 236, 238.

10. Michael I. Pupin and Edwin Howard Armstrong, "Radioreceiving System

Having High Selectivity," U.S. Patent No. 1,416,061, application date: 18 Dec. 1918, issue date: 16 May 1922.

11. Edwin Howard Armstrong, "Wave Signaling System," U.S. Patent No. 1,716,573, application date: 24 Feb. 1922, issue date: 11 June 1929.

12. Edwin Howard Armstrong, "Radio Signaling System," U.S. Patent No. 2,082,935, application date: 6 Aug. 1927, issue date: 8 June 1937; and, "Methods of Reducing the Effects of Atmospheric Disturbance," *PIRE* 16 (Jan. 1928): 15–26.

13. John Renshaw Carson, "The Reduction of Atmospheric Disturbances," *PIRE* 16 (July 1928): 966–75, quotations on pp. 966, 974, 966.

14. Edwin H. Armstrong, "Radio Telephone Signaling," U.S. Patent No. 1,941,447, application date: 18 May 1927, issue date: 26 Dec., 1933.

15. Ibid.

16. Edwin Howard Armstrong, "Radio Signaling System," U.S. Patent 1,941,066, application date: 30 July 1930, issue date: 26 Dec. 1933.

17. Murray G. Crosby, "Phase Modulation Receiver," U.S. Patent No. 2,101,703, application date: 23 Jan. 1932, issue date: 7 Dec. 1937, assigned to RCA.

18. Ellison S. Purington, "Fortune Magazine, October, 1939," memorandum, 25 Oct. 1939, box 275, microfilm reel, "RCA-1 C.S. & M: #1 to #1419," frame 364, AP.

19. Edwin H. Armstrong, "Radiosignaling," U.S. Patent No. 1,941,068, application date: 24 Jan. 1933, issue date: 26 Dec. 1933; and "Radiosignaling," U.S. Patent No. 1,941,069, application date: 24 Jan. 1933, issue date: 26 Dec. 1933.

20. Armstrong, U.S. Patent No. 1,941,069.

21. The FCC can more easily regulate broadcast FM than AM for two technical reasons. First, FM operates in a part of the spectrum where propagation is limited to a few miles beyond "line of sight." Also, FM receivers sort out overlapping signals far better than do AM receivers.

22. Armstrong, U.S. Patent No. 1,941,068.

23. Ibid.

24. Armstrong, U.S. Patent No. 1,941,069.

25. Ibid. (emphasis added).

26. Armstrong, U.S. Patent No. 1,941,447.

27. Armstrong, U.S. Patent No. 1,941,069.

28. Ibid.

29. Ibid. (emphasis added).

30. Ibid.

31. J. C. Johnson, "Thermal Agitation of Electricity in Conductors," *Physical Review* 32 (July 1928): 97–109.

32. Armstrong, U.S. Patent No. 1,941,069.

33. Edwin Howard Armstrong, "Demonstration of Reduction of Tube Noise by Frequency Modulation at 7.5 meters," hand-drawn sketch, 21 July 1932, box 159, AP.

34. Armstrong to Moses & Nolte, 23 Aug. 1932, box 245, AP; Armstrong, U.S. Patent No. 1,941,069. Moses and Nolte were Armstrong's patent lawyers.

35. Armstrong, U.S. Patent No. 1,941,069.

36. Ibid.

37. Armstrong to Gano Dunn, 22 June 1939, box 163, AP. In 1935 Armstrong stated that FM's first demonstration "was early in 1934 to a group of scientists." "Radio Device Ending Fading, Static Reported," *New York Herald Tribune*, 26 Apr. 1935.

38. Harold H. Beverage, "History of Frequency Modulation Development in R.C.A. Communications, Inc.," memorandum, 27 Oct. 1939, box 275, microfilm reel, "RCA-1 C.S. & M: #1 to #1419," frame 386, AP.

39. Crosby, "Crosby Notebook," 5 Jan. 1934, box 160, manila envelope, "Notes from Laporte's Original examination," AP.

40. Beverage to Taylor, 15 June 1934, box 275, microfilm reel, "Cravath, Swaine & Moore Folder #190M: Index 90 to #5922," frame 5205, AP.

41. For descriptions of the Empire State Building Television Laboratory FM tests, see Edwin Howard Armstrong, "A Method of Reducing Disturbances in Radio Signaling by a System of Frequency Modulation," *PIRE* 24 (May 1936): 717–40; Thomas J. Buzalski, "Field Test of the Armstrong Wide-Band Frequency Modulation System from the Empire State Building, 1934 and 1935" *A.W.A. Review* 1 (1986): 109–16, reprinted in Morrisey, *The Legacies of Edwin Howard Armstrong*, 244–50.

42. Radio Corporation of America, *Annual Report, 1933* (New York: RCA, 1934), 6.

43. Beverage to Charles J. Young, 8 Apr. 1932, box 275, microfilm reel, "RCA-7 C.S. & M: 191-M-5923 to 226-M-7212," frame 5972, AP.

44. Record of visitors to Empire State Building, 12 Jan. 1934, cited by Buzalski, "Field Test of the Armstrong Wide-Band Frequency Modulation System," 110.

45. Beverage to Taylor, 15 June 1934.

46. In 1920 Armstrong had sold to Westinghouse his patent rights to feedback and two other inventions for an immediate payment of $350,000, and—if the courts ruled in favor of Armstrong—an additional $200,000.

47. Lessing, *Man of High Fidelity*, 51, 130–31.

48. Ibid., esp. 158–93.

49. Quoted in ibid., 189.

50. Ibid., 190.

51. R. E. Shelby, "Monthly Report, March 1934," memorandum, 31 Mar. 1934, box 275, microfilm reel, "RCA-1 C.S. & M: #3436 to #4696," frame 3925, AP.

52. Armstrong, "Method for Reducing Disturbances," 717–18.

53. Shelby, "Monthly Report, March 1934."

54. Buzalski, "Field Tests of the Armstrong Wide-Band Frequency Modulation System," 113.

55. Ibid.; R. E. Shelby, "Monthly Report: Empire State Television Laboratory," memorandum, 31 May 1934, box 275, microfilm reel, "RCA-1 C.S. & M: #3436 to #4696," frame 3927, AP.

56. Tunick to Mr. Martin, 23 May 1934, box 164, envelope, "Miscellaneous RCA Notes and Records," AP.

57. Armstrong, "Method of Reducing Disturbances," 718.

58. George E. Burghard, "Eighteen Years of Amateur Radio," *Radio Broadcast* 2 (Aug. 1923): 290.

59. Hansell to Beverage, 20 June 1934, box 275, microfilm reel, "RCA-1 C.S. & M: #1 to #1419," frames 12 and 400, AP.

60. Lessing, *Man of High Fidelity*, 221.

61. Armstrong, "Method of Reducing Disturbances," 720.

62. Beverage to Taylor, 15 June 1934.

63. Ibid.

64. Hansell to Beverage, 20 June 1934.

65. Ibid.

66. Ibid.

67. R. E. Shelby, "Monthly Report," memorandum, 2 July 1934, box 275, microfilm reel, "RCA-1 C.S. & M: #3436 to #4696," frame 3928, AP.

68. Sadenwater to Wozencraft, 15 Mar. 1940, box 159, AP.

69. Beverage to Beal, 28 July 1935, box 275, microfilm reel, "Cravath, Swaine & Moore Folder #190M: Index 90 to #5922," frame 5104, AP.

70. Hansell to Beverage, 29 July 1935, box 275, microfilm reel, "Cravath, Swaine & Moore Folder #190M: Index 90 to #5922," frame 5410, AP.

71. Ibid.

72. Charles M. Burrill, "Status Report for Weeks of October 15 and 22, 1934," memorandum, 22 Oct. 1934, box 159, AP.

73. Charles M. Burrill, "Status Report for Weeks of Oct. 29 and Nov. 5, 1934," memorandum, 7 Nov. 1934, box 159, AP.

74. Charles M. Burrill, "Notes on the Reception of Amplitude and Wide-Band Frequency Modulated Signals from the Empire State Transmitter, Dec. 12 to 14, 1934, at 22 Mountain View Rd., Millburn, N.J.," memorandum, 23 Jan. 1935, box 159, AP.

75. Lessing, *Man of High Fidelity*, 226.

76. Ibid., 225. To date, a copy of this letter has not been found.

77. FCC, *Report on Chain Broadcasting* (Washington, D.C.: Government Printing Office, 1941), 18–20.

78. *Radio Corporation of America, Seventeenth Annual Report: Radio Corporation of America, Year Ended December 31, 1936* (New York: RCA, 1937), 11–12.

79. FCC, *Report on Chain Broadcasting*, 14–15.

80. Ibid., 18–20. This part of the report described the wide-ranging subsidiaries of RCA, warning of "broader problems raised by this concentration of power in the hands of a single group."

81. Elmer Engstrom to Baker, 12 March 1935, box 275, microfilm reel, "RCA-1 C.S. & M: #1 to #1419," frame 438, AP.

82. Baker to Schairer, 19 Mar. 1935, box 183, and box 164, envelope labeled "Miscellaneous RCA Notes and Records," AP.

83. Beal to Schairer, 14 May 1935, box 275, microfilm reel, "RCA-1 C.S. & M: #1 to #1419," frame 227, AP.

84. Beal, "Comments on the Characteristics of the Armstrong Wide Band Frequency Modulation System and Suggestions for Field Tests Related to Extension of Field Test Area," memorandum, 9 July 1935, box 275, microfilm reel, "Cravath, Swaine & Moore Folder #190M: Index 90 to #5922," frame 5415, AP.

85. "Radio Device Ending Fading, Static Reported: Armstrong Perfects Method Increasing Range of Ultra-Short Wave Broadcasting," *New York Herald Tribune*, 26 Apr. 1935.

86. "Radio Invention Helps End Static," *New York Times*, 26 Apr. 1935.

87. "Major Armstrong Fights Static," *Electronics* 8 (May 1935): 162.

88. "Events of the Future Foretold," *Electronics* 8 (June 1935): 203.

89. "A Treatise on Frequency Modulation," *Communication and Broadcast Engineering* 2 (June 1935): 18; "Technical Topics: Frequency Modulation: Major Armstrong," *QST* 19 (Sept. 1935): 21.

90. Armstrong to Beal, 5 Aug. 1935, box 275, microfilm reel, "RCA-1 C.S. & M: #1 to #1419," frame 540, AP.

91. Baker to Armstrong, 5 Aug. 1935, box 275, microfilm reel, "RCA-1 C.S. & M: #1 to #1419," frame 444, AP.

92. Armstrong to Beal, 12 Aug. 1935, box 275, microfilm reel, "RCA-1 C.S. & M: #1 to #1419," frame 496, AP.

93. Harold H. Beverage, Harold O. Peterson, Murray G. Crosby, Bertram Trevor, and Charles Burrill, "F-18-7: Wide Band Frequency Modulation Tests," memorandum, 9 Oct. 1935, box 275, microfilm reel, "Cravath, Swaine & Moore Folder #190M: Index 90 to #5922," frame 5862, AP.

Chapter Five • FM Pioneers, RCA, and the Reshaping of Wideband FM Radio, 1935–1940

Epigraphs. [Lawrence P. Lessing], "Revolution in Radio," *Fortune* 20 (Oct. 1939): 86. Ellison S. Purington, "Fortune Magazine, October, 1939," memorandum, 25 Oct. 1939, box 275, microfilm reel, "RCA-1 C.S. & M: #1 to #1419," frame 364, AP.

1. Young to Armstrong, 1 Nov. 1935, box 162, envelope, "Conklin Setters; also RCA etc. Tunick & Sadenwater," AP; Armstrong to Young, 6 Nov. 1935, box 162, envelope, "Conklin Setters; also RCA etc. Tunick & Sadenwater," AP.

2. Beal to Schairer, 7 Nov. 1935, box 275, microfilm reel, "RCA-1 C.S. & M: #1 to #1419," frame 229, AP.

3. Donald G. Fink, "Phase-Frequency Modulation," *Electronics* 8 (Nov. 1935): 17; "Armstrong Demonstrates Frequency Modulation System," *Communication and Broadcast Engineering*, Nov. 1935, 21.

4. Lawrence P. Lessing, *Man of High Fidelity: Edwin Howard Armstrong* (Philadelphia: J. B. Lippincott, 1956), 209–10.

5. "Editorial: High-Quality Programs," *Communication and Broadcast Engineering*, Nov. 1935, 4.

6. Edwin H. Armstrong, "Radiosignaling," U.S. Patent No. 1,941,068, application date: 24 Jan. 1933, issue date: 26 Dec. 1933.

7. Thomas J. Buzalski, "Field Tests of the Armstrong Wide-Band Frequency Modulation System from the Empire State Building, 1934 and 1935," *A.W.A. Review* 1 (1986): 109–16, reprinted in John W. Morrisey, ed., *The Legacies of Edwin Howard Armstrong* (n.p.: Radio Club of America, 1990), 244–50.

8. Beal to Schairer, 7 Nov. 1935.

9. Gleason L. Archer, *History of Radio to 1926* (New York: American Historical Society, 1938), 284–87; FCC, *Report on Chain Broadcasting* (Washington, D.C.: Government Printing Office, 1941), 5–6.

10. "Regionals Form Organization with Shepard Named President: Permanent Setup Adopted at Meeting in Chicago; Advocate 5,000 Watts Both Night and Day," *BBA* 14 (15 May 1938): 14.

11. See Susan Smulyan, *Selling Radio: The Commercialization of American Broadcasting, 1920–1934* (Washington, D.C.: Smithsonian Institution Press, 1994); and Robert W. McChesney, *Telecommunications, Mass Media, and Democracy: The Battle for the Control of U.S. Broadcasting, 1928–1935* (New York: Oxford Press, 1993).

12. This is the central argument in Charles A. Siepmann's book, *Radio's Second Chance* (Boston: Atlantic-Little, Brown Books, 1946). For somewhat different, largely commercial, reasons, editor Milton B. Sleeper also called FM "radio's second chance" in "Revolution for Profit," *FM* 1 (Nov. 1940): 3.

13. [Lawrence P. Lessing], "Revolution in Radio," *Fortune* 20 (Oct. 1939); Alfred Toombs, "The Radio Battle of 1941: FM vs. AM," *Radio News*, Mar. 1941, 7; Harry Sadenwater to F. R. Deakins, 15 May 1936, box 275, microfilm reel, "RCA-1 C.S. & M: #1 to #1419," frame 551, AP.

14. Beal to Schairer, 13 Dec. 1937, box 275, microfilm reel, "RCA-1 C.S. & M: #1 to #1419," frame 565, AP.

15. Bruce Robertson, "Armstrong Soon to Start Staticless Radio: Broad Claims for New System Are Made," *BBA* (1 Feb. 1939): 19.

16. "Yankee Frequency Modulation About Ready: Armstrong Method to Go on Air in June," *BBA* 16 (1 June 1939): 19.

17. Armstrong to Andrew Ring, 27 Apr. 1936, box 423, AP; "High Power Frequency Modulation," *Electronics* 9 (May 1936): 25; "Maj. Armstrong Granted CP for 40 kw. Apex Test," *BBA* 10 (15 June 1936): 51.

18. Edwin Howard Armstrong, "Application for Modification of Radio Station Construction Permit," 24 Apr. 1936, box 5, AP.

19. Donald G. Fink, "From the Mountaintops," *Technology Review* 41 (Apr. 1939): 257.

20. Donald G. Fink, "FM Gets Its 'Day in Court,'" *Electronics* 13 (Apr. 1940): 14.

21. Harry Sadenwater to C. K. Throckmorton, memorandum, 16 Mar. 1938, box 159, AP.

22. "High-Fidelity Signals Free from Static Are Shown in Tests by Maj. Armstrong," *BBA* 16 (1 Apr. 1939): 81.

23. Henry M. Lane, "Engineers Hail Noiseless Radio," *Boston Sunday Post*, 28 May 1939. The author thanks Donna Halper for sharing this article. That Lessing lists identical high-fidelity sound effects—for example, tearing paper, pouring water, and the striking of a bell—in his similarly worded description of the November 1935 IRE demonstration suggests that he used this article as a source for his description of the IRE event.

24. Ibid.

25. To clarify, FM receivers can sort out two incoming signals if one signal is at least 3 dB stronger than the other. In other words, the power ratio of the two signals must be at least 2.0. Bruce Robertson, "Armstrong Soon to Start Staticless Radio: Broad Claims for New System Are Made," *BBA* 16 (1 Feb. 1939): 19; "High-Fidelity Signals Free from Static Are Shown in Tests by Maj. Armstrong," *BBA* 16 (1 Apr. 1939): 81; Irwin R. Weir, "Field Tests of Frequency- and Amplitude-Modulation with Ultrahigh-Frequency Waves, Part I," *General Electric Review* 42 (May 1939): 188–91; Irwin R. Weir, "Field Tests of Frequency- and Amplitude-Modulation with Ultrahigh-Frequency Waves, Part II," *General Electric Review* 42 (June 1939): 270–73.

26. "High-Fidelity Signals Free from Static Are Shown in Tests by Maj. Armstrong," 81.

27. "Half-Million Will Be Spent for Tests: Shepard Starts 50 kw. Plant for Armstrong Experiments," *BBA* 14 (15 Jan. 1938): 15.

28. The 5,000-cycle limitation for long-distance telephone lines was mentioned in an anonymous internal RCA memorandum: "Transmitter Advanced Development Section, Radio Corporation of America. Frequency Modulation: A brief discussion of its principles of operation, the claimed advantages and disadvantages, activities outside of and within RCA companies, and engineering work to be done," Dec. 1939, box 159, AP.

29. Henry M. Lane, *Boston Sunday Post*, 7 Jan. 1940, cited by "First F-M Network Broadcast: Yonkers Program Received in Boston through Use of Four Experimental Transmitters," *BBA* 18 (15 Jan. 1940): 32.

30. "F-M Broadcasting on Three Relays Proves Successful: Armstrong and Doolittle See Widespread Radio Changes," *BBA* 17 (15 Dec. 1939): 26.

31. "Broadcasters Organize Group for Operation of F-M Stations: Charter Sought after New York Meeting as Plans Are Laid; Shepard Elected Chairman," *BBA* 18 (15 Jan. 1940): 31.

32. Ibid.

33. "Yankee Asks FCC for Regular License for 50 kw. F-M Station in New York," *BBA* 17 (1 Nov. 1939): 64.

34. "Jett Orders F-M Study," *BBA* 17 (1 Dec. 1939): 77.

35. The date of the hearing was originally scheduled for 28 February, but at Armstrong's request the FCC postponed the meeting by three weeks. "FCC to Investigate Progress of FM: All Phases of New Art Will Be Probed at Hearing," *BBA* 18 (1 Jan. 1940): 19. The FCC's official notice was reprinted in "Agenda of Feb. 28 Hearing on Frequency Modulation," *BBA* 18 (1 Jan. 1940): 19; "Armstrong Asks Delay Pending F-M Study," *BBA* 18 (1 Feb. 1940): 70.

36. Murray G. Crosby, "Frequency Modulation Propagation Characteristics," *PIRE* 24 (June 1936): 898–913; "Frequency Modulation Noise Characteristics," *PIRE* 25 (Apr. 1937): 472–514; "Carrier and Side-Frequency Relations with Multi-Tone Frequency or Phase Modulation," *RCA Review* 3 (July 1938): 103–7; "Communication by Phase Modulation," *PIRE* 27 (Feb. 1939): 126–36; and "The Service Range of Frequency Modulation," *RCA Review* 4 (Jan. 1940): 349–71.

37. Baker to Armstrong, 29 Nov. 1937, box 244, AP.

38. Sadenwater to Wozencraft, 15 Mar. 1940, box 159, AP.

39. Beal to Schairer, 13 Dec. 1937.

40. Hansell to Armstrong, 2 June 1936, box 183, AP.

41. Sadenwater to Deakins, 15 May 1936.

42. Sadenwater to J. L. Schwank, 13 Jan. 1937, box 183, AP.

43. Sadenwater to Throckmorton, 16 Mar. 1938.

44. Ibid.

45. Ibid.

46. Dale Pollack, "Suggestions for Future Development Projects," memorandum, 13 Sept. 1937, box 163, AP.

47. For a biography of Baker, see "Walter R. G. Baker: 1892–1960," Web page, www.ieee.org/web/aboutus/history_center/biography/baker.html, viewed 3 Apr. 2009.

48. Dale Pollack, "Report on Trip to the Convention: Rochester, November, 1938," memorandum, 21 Nov. 1938, box 163, AP. Also, box 183, AP.

49. Dale Pollack, "Frequency Modulation Development Program," memorandum, 22 Nov. 1938, box 163, AP.

50. O. B. Hanson to Lenox R. Lohr, 18 Jan. 1939, box 275, microfilm reel, "No. RCA-2, first original #1420, last original #3089," frame 2350, AP.

51. Pollack, "Frequency Modulation Development Program."

52. Hansell to Niles Trammell, 25 May 1939, box 244, AP.

53. Hanson to Lohr, 24 Jan. 1939, box 275, microfilm reel, "No. RCA-2, first original #1420, last original #3089," frame 2318, AP.

54. Ibid.

55. "Frequency Modulation: A Revolution in Broadcasting?" *Electronics* 13 (Jan. 1940): 10.

56. F. R. Deakins to Lewis M. Clement, 28 Sept. 1939, box 183, AP.

57. Hansell to Trammell, 25 May 1939.

58. Trammell to Hanson, 31 May 1939, box 244, AP.

59. Beal to Clement, 15 May 1939, box 275, microfilm reel, "RCA-1 C.S. & M: #3436 to #4696," frame 4447, AP.

60. R. D. Duncan Jr. to Clement and Pollack, 9 June 1939, box 183, AP.

61. Hanson to Lohr, 28 Aug. 1939, box 275, microfilm reel, "RCA-4 C.S. & M: #3435 to #4696," frame 3520, AP.

62. Hanson to Lohr, 15 Sept. 1939, box 275, microfilm reel, "RCA-4 C.S. & M: #3435 to #4696," frame 3522. Also, reel #2, frame 2358, AP.

63. See, for example: [Lessing], "Revolution in Radio"; "Armstrong's Threat to Upset Radio Applecart Marked by FM-Television Battle for Bands," *Newsweek*, 1 Apr. 1940: 30.

64. Orrin E. Dunlap Jr., "Divided Opinions," *New York Times*, 17 Mar. 1940; Lewis V. Gilpin and Sol Taishoff, "Birth of Commercial FM This Year Seen," *BBA* (1 Apr. 1940): 18.

65. Gilpin and Taishoff, "Birth of Commercial FM This Year Seen," 19.

66. Fink, "FM Gets Its 'Day in Court,'" 15.

67. Gilpin and Taishoff, "Birth of Commercial FM This Year Seen."

68. Fink, "FM Gets Its 'Day in Court.'"

69. Gilpin and Taishoff, "Birth of Commercial FM This Year Seen."

70. Ibid.

71. FCC, "In the Matter of Aural Broadcasting on Frequencies above 25,000 kilocycles Particularly Relating to Frequency Modulation," Docket No. 5805, 20 May 1940. Document available from FCC Web page, www.fcc.gov.

Conclusion

Epigraphs. John R. Poppele, "FM and Its Economic Advantages," *Proceedings of the Radio Club of America* 17 (Oct. 1940): 11–12. Robert Angus, "What's Wrong with American FM?" *Popular Electronics* 16 (June 1962): 45.

1. John Law, "Technology, Closure, and Heterogeneous Engineering: The Case of the Portuguese Expansion" in Wiebe E. Bijker, Thomas Parke Hughes, and Trevor J. Pinch, eds., *The Social Construction of Technological Systems: New Directions in the Sociology and History of Technology* (Cambridge, Mass.: MIT Press, 1987), 111–34.

2. Boelie Elzen, Bert Enserink, and Wim A. Smit, "Socio-Technical Networks: How a Technology Studies Approach May Help to Solve Problems Related to Technological Change," *Social Studies of Science* 26, no. 1 (Feb. 1996): 95–141.

3. "FM to Establish New York Office; Dorrance Named," *BBA* 18 (1 June 1940): 14.

4. Donald MacKenzie and Judy Wajcman, "Introductory Essay: The Social Shaping of Technology," in Mackenzie and Wajcman, eds., *The Social Shaping of Technology*, 2d ed. (Buckingham: Open University Press, 1999), 3–27, quotation on p. 4.

5. Thomas Jay Misa, "Retrieving Sociotechnical Change from Technological Determinism," in Merritt Roe Smith and Leo Marx, eds., *Does Technology Drive History? The Dilemma of Technological Determinism* (Cambridge, Mass.: MIT Press, 1994), 115–42, quotation on p. 117.

6. Walter G. Vincenti, "The Technical Shaping of Technology: Real-World Constraints and Technical Logic in Edison's Electrical Lighting System," *Social Studies of Science* 25, no. 3 (Aug. 1995): 553–74, quotation on p. 556.

7. For more about technologies with "embedded" politics, see Langdon Winner, "Do Artifacts Have Politics?" *Daedalus* 109 (Winter 1980): 121–36.

8. The commercial schedule was reported as "Operating power: 250 w. or less, $300; 1 kw., $500; 2 kw., $750; 5 kw., $1,250; 10 kw., $2,000; 20 kw., $3,000; 30 kw., $3,750; 40 kw., $4,500; 50 kw. or more, $5,000 for 50 kw.; $50 for each additional kw." See "Armstrong Fixes Royalty Payments," *BBA* 18 (15 Jan. 1940): 31.

9. See Charles A. Siepmann, *Radio's Second Chance* (Boston: Atlantic-Little, Brown Books, 1946).

10. "Finally, FCC Okays Stereo," *Broadcasting* 60 (19 Apr. 1961): 65–66. The FCC considered several methods of implementing stereophonic broadcasts, including one by Murray Crosby. The commission chose a design developed jointly by General Electric and Zenith.

11. Christopher H. Sterling and John Michael Kittross, *Stay Tuned: A Concise His-*

tory of American Broadcasting (Mahwah, N.J.: Lawrence Erlbaum Associates, 2001), 828.

12. Angus, "What's Wrong with American FM?"

13. Ibid.

14. Woody Allen, director, *Annie Hall*, distributed by United Artists, 1977.

15. Susan J. Douglas, *Listening In: Radio and the American Imagination, from Amos 'n' Andy and Edward R. Murrow to Wolfman Jack and Howard Stern* (New York: Times Books, 1999), 256–84; Christopher H. Sterling and Michael C. Keith, *Sounds of Change: A History of FM Broadcasting in America* (Chapel Hill: University of North Carolina Press, 2008).

16. "FM Signals Follow Several Horizons, Armstrong Tells Indiana Radio Session," *BBA* 19 (1 July 1940): 52. It should be noted that Armstrong made this point to support the dubious claim that the past teaches that "everything that has been accomplished in science was at one time sworn to be impossible."

17. John R. Poppele, "FM and Its Economic Advantages," *Proceedings of the Radio Club of America* 17 (Oct. 1940): 11–12.

18. For information about low-power FM, see the Prometheus Radio Project Web page, www.prometheusradio.org, viewed 3 April 2009. For FCC documents about low-power FM, see the FCC Web page, www.fcc.gov.

GLOSSARY

amplitude modulation (AM). A method of encoding a carrier wave to convey information. The amplitude of the carrier is varied (modulated) according to the amplitude of an audio-frequency wave.

antenna. A metallic apparatus for sending or receiving electromagnetic waves.

arc oscillator. An early continuous wave radio transmitter.

audio. Of or relating to the transmission, reception, or reproduction of sound.

audio amplifier. An electronic device that increases the amplitude of reproduced sound. Audio amplifiers are often subsystems of radio transmitters and receivers.

audion. The first electronic amplifier, a grid triode invented by Lee de Forest in 1906. The audion was the ancestor of the vacuum tube and the transistor.

balanced amplifier. A symmetrically structured amplifier with two branches having identical or nearly identical properties. Often used to subtract or add two signals.

bandwidth. The numerical difference between the upper and lower frequencies of a band of frequencies. The audio bandwidth of an amplitude-modulation system is approximately one-half the width of the radio channel.

binary amplitude modulation. A means for modulating an electromagnetic wave to carry telegraph messages.

breadboard. An experimental prototype of an electric circuit or system, often mounted on a perforated board.

capacitor. See condenser.

carrier (or carrier wave). An electromagnetic wave that can be modulated, as in frequency, amplitude, or phase, to transmit speech, music, images, or other signals.

cascaded amplifiers. A circuit in which the output of an amplifier is connected to the input of a succeeding amplifier.

channel. A specified radio-frequency band for the transmission and reception of electromagnetic signals, as for radio or television signals.

coherer. A device once used to detect electromagnetic waves in a wireless (radio) signaling system.

condenser (or capacitor) (symbol C). A capacitive circuit element that blocks electric current and holds a charge. Often connected to an inductor to form a resonant circuit.

continuous wave. A sinusoidal wave of constant amplitude and frequency.

crystal detector. A rectifying detector used especially in early radio receivers and consisting of a semiconductor crystal in point contact with a fine metal wire.

damped wave. An oscillating wave whose amplitude decays to zero.

detection (also demodulation). The extraction of sound waves from a modulated carrier wave.

detector (also demodulator). A device that extracts sound waves from a modulated radio carrier wave.

electromagnetic waves. Energy comprising electrical and magnetic components. Radio waves are electromagnetic waves traveling through space.

electronic. Describes devices that are based on the control of electron flow. During the first half of the twentieth century, almost all electronic devices used vacuum tubes.

facsimile (fax). To transmit an image by electronic means.

fading. Fluctuation in the strength of incoming radio signals, usually due to changing atmospheric conditions.

fidelity. The degree to which an electronic system accurately reproduces sound.

frequency. The number of cycles of a waveform per second.

frequency deviation. In FM, the amount of frequency shift above or below the unmodulated carrier. The frequency deviation is one-half of the frequency swing.

frequency modulation (FM). A method of encoding a carrier wave to convey information. The frequency of the carrier is varied according to the amplitude of an audio-frequency wave.

frequency multiplier. An electronic device that multiplies the frequency of an input signal, usually by a factor of two or three. In FM, a frequency multiplier is used to multiply the frequency deviation of a modulated carrier wave.

frequency-shift keying (FSK). The use of frequency modulation to transmit digital data, usually by Morse code or similar telegraph code messages.

frequency swing. In FM, twice the frequency deviation.

heterodyne. An electrical or electronic circuit that combines two radio-frequency waves in order to produce a new wave that is either the sum or the difference of the frequencies of the original waves.

high fidelity ("hi-fi"). The electronic reproduction of sound with minimal distortion and wide frequency response.

inductance (symbol L*).* A circuit element, typically a conducting coil.

interference. Degradation of reception on account of electromagnetic noise or undesired signals.

intermediate frequency (IF). The fixed frequency of the middle stage (i.e., IF amplifier) of a superheterodyne radio receiver. Most of the overall amplification that takes place in a receiver occurs in the IF amplifier stage.

kilocycles per second (also kilohertz). A unit of frequency equal to 1,000 hertz, or 1,000 cps.

LC *circuit.* A resonant circuit composed of an inductive element (*L*) and a capacitive element (*C*), and which is used for tuning.

megacycles per second (also kilohertz). A unit of frequency equal to 1,000,000 hertz, or 1,000,000 cps.

modulation. The variation of the amplitude, frequency, or phase, of a carrier wave.

Morse code. Either of two codes used for transmitting messages in which letters of the alphabet and numbers are represented by various sequences of dots (short marks), dashes (long marks), and spaces. The letter *A*, for example, is represented by the American Morse code with a dot-dash sequence: • — .

multipath fading. Fading in reception when the transmitted signal propagates via two paths of different lengths. The difference creates a relative shift in phase, thereby causing the otherwise identical signals partially to cancel each other out.

narrowband. Responding to or operating at a narrow band of frequencies.

narrowband FM. Traditionally refers to FM systems with a channel width of less than 10,000 cps.

phase modulation. A method of encoding a carrier wave to convey information. The phase of the carrier is varied according to the amplitude of an audio-frequency wave.

propagation. The process by which electromagnetic waves are transmitted though a medium, such as air or free space.

radiotelephone receiver. A device that receives incoming modulated radio signals and converts them to sound waves.

radiotelephone transmitter. A device that generates and amplifies a carrier wave, modulates it with a sound wave, and radiates the resulting wave with an antenna.

rectify. To convert alternating current into direct current.

resonant circuit. An electric circuit that is tuned to allow the greatest flow of current at a certain frequency. The most common types of resonant circuits are composed of reactive elements (*LC*) or crystals.

sideband. Either of the two bands of frequencies, one just above and one just below a carrier frequency, that result from modulation of a carrier wave.

slope detector. A simple detector of frequency-modulation waves, based on the sloped response of an *LC* circuit.

spark gap. A device once used to transmit wireless messages.

spectrum. The distribution of energy emitted by a radiant source, as by an incandescent body, arranged in order of frequencies.

static. Random radio noise caused by atmospheric disturbances or man-made electrical interference.

superheterodyne. 1. An electronic version of the heterodyne circuit in which an incoming radio signal is combined with a locally generated continuous wave to produce a standard intermediate frequency. Superheterodyne circuits are used to simplify amplification and tuning. 2. A radio receiver designed with a superheterodyne circuit.

tube hiss. The molecular- or quantum-level white noise produced by vacuum tubes.

tune. 1. To adjust a transmitter, receiver, or circuit to reject or accept a band of radio waves. 2. To adjust a resonant circuit to oscillate at a single frequency.

vacuum tube. An electron tube from which all or most of the gas has been removed. Vacuum tubes are typically used for the electronic amplification or rectification of electric waves.

wavelength. The distance between succeeding crests of a sound wave, electrical wave, or radio wave. For radio waves, the wavelength equals the speed of light divided by the frequency.

wideband. Responding to or operating at a wide band of frequencies.

wideband FM. Traditionally synonymous with Armstrong FM. Something of a misnomer, the adjective "wide" refers to the channel width, 200,000 cps.

Almost all literature about the history of frequency modulation before World War II echoes the narrative of Lawrence Lessing's hagiographic biography, *Man of High Fidelity: Edwin Howard Armstrong* (Philadelphia: J. B. Lippincott, 1956; 2d ed., New York: Bantam Books, 1969). Lessing's book exhibits a number of glaring errors and distortions, though. By focusing almost exclusively on Armstrong, it all but ignores the work of anyone else, such as when it dismisses the Poulsen arc merely as an "unsuccessful attempt to employ [frequency modulation]" (p. 198). Moreover, it falsely implies that RCA and AT&T researchers labored mostly with narrowband frequency modulation to make FM broadcast radio practical.

The books that more or less follow Lessing vary widely in quality in how they tell the history of frequency modulation. Don V. Erikson's *Armstrong's Fight for FM Broadcasting: One Man vs. Big Business and Bureaucracy* (University: University of Alabama Press, 1973) is based almost entirely on sources that Lessing cites. Tom Lewis's *Empire of the Air: The Men Who Made Radio* (New York: Edward Burlingame Books, 1991) is more useful and original, in large part because Lewis used material in the Armstrong Papers that Lessing and Erikson apparently neglected, and because Lewis acknowledges the general incompleteness of Lessing's narrative. Christopher H. Sterling and Michael C. Keith's *Sounds of Change: A History of FM Broadcasting in America* (Chapel Hill: University of North Carolina Press, 2008) does not diverge much from Lessing's narrative about prewar frequency modulation, but it is the best history of postwar FM broadcasting. Hugh R. Slotten's *Radio and Television Regulation: Broadcast Technology in the United States, 1920–1960* (Baltimore: Johns Hopkins University Press, 2000) contains a chapter about postwar FM radio that is based principally on government documents and credibly disputes Lessing's interpretation. See "'Rainbow in the Sky': FM Radio, Technological Superiority, and Regulatory Decision Making, 1936–1948" (pp. 113–44). Researchers should also consult an anthology published by the Radio Club of America, John W. Morrisey, ed., *The Legacies of Edwin Howard Armstrong* (n.p.: Radio Club of America, 1990), which offers several first-person articles written by men who witnessed the development of early FM broadcasting.

By far the most significant primary source for this book was the Edwin Howard Armstrong Papers collection, located in the Rare Books and Manuscripts Library of Columbia University. A large portion of this collection, comprising more than five hundred boxes, including dozens of microfilm reels, is made up of material that originated within RCA. During a lawsuit that began in 1948, the law firm that represented Armstrong obtained copies of every RCA document related to frequency modulation, including correspondence, research reports, and sales literature. In 1990 that law firm donated these documents to the Armstrong Papers collection and thus researchers have at their disposal an archive of frequency-modulation work within the RCA organization that Lessing did not.

This study also depended on technical literature for primary sources. Researchers should consult the appendix to this book, which lists all patents applied for before 1941 that related to FM radio. These were culled largely from memoranda written by RCA engineers, managers, and patent lawyers. Any one researching FM's history should take care to read patents with no preconceptions and not be led astray by Lessing. For more than half a century, almost every historian of FM radio has accepted his description of what he calls "Armstrong's [four] basic patents of 1933" (p. 205). But only two of these patents describe a wideband FM system, and rather than claim that FM radio suppresses static, one implied that frequency modulation has no effect on static.

Because radio technology evolved so quickly during the early twentieth century, books tended to be less useful primary sources of technical literature for this study than periodicals did, although almost nothing was published about frequency modulation before 1934. The titles of all the magazines and journals cited in this book are too numerous to list, but the most valuable and sophisticated articles appeared in the *Proceedings of the Institute of Radio Engineers.* See also *Proceedings of the Radio Club of America, Communication and Broadcast Engineering,* and *Electronics,* all of which also printed articles about the history of frequency modulation. Articles about FM technology and the broadcasting industry written for the layperson were published in various magazines and newspapers, including the most important trade magazine of the 1930s and 1940s, *Broadcasting-Broadcasting Advertising,* as well as the *New York Times,* the *New York Herald-Tribune,* and the British magazine *Wireless World.* The earliest issues of *FM* magazine, which debuted in November 1940, contained many articles about the history of frequency modulation. A helpful anthology of technical articles about early radio is George Shiers, ed., *The Development of Wireless to 1920* (New York: Arno Press, 1977).

This book depended on many government documents. Researchers should read the annual reports of the Federal Radio Commission and of the Federal Communications Commission (Washington, D.C.: Government Printing Office). In 1948 two congressional committees investigated the history of frequency-modulation radio and produced reports that provide valuable transcripts of testimony of dozens of individuals who pioneered frequency-modulation broadcasting. See the two-part report of the House Committee on Interstate and Foreign Commerce, *Radio Frequency Modulation: Hearings on H. J. Res. 78: A Joint Resolution Relating to Assignment of a Section of the 50-Megacycle Band of Radio Frequencies for Frequency Modulation (FM),* 80th Cong., 2d sess., 1948; as well as the Senate Committee on Interstate and Foreign Commerce, *Progress of FM Radio: Hearings on Certain Charges Involving Development of FM Radio and RCA Parent Policies,* 80th Cong., 2d sess., 1948. During the 1940s, the FCC commissioned a study of chain (i.e., network) broadcasting. It was published as FCC, *Report on Chain Broadcasting* (Washington, D.C.: Government Printing Office, 1941), and was reprinted in Christopher H. Sterling, ed., *Special Reports on American Broadcasting, 1932–1947* (New York: Arno Press, 1974). Five years later, the report's author, Charles A. Siepmann, published *Radio's Second Chance* (Boston: Atlantic-Little, Brown Books, 1946), which argues that FM radio could be an antidote to errors committed by the FRC and FCC in the regulation of AM broadcasting. For primary documents about the regulation of radio before 1927, the *Radio Service Bulletin*, issued monthly by the U.S. Commerce Department's Bureau of Navigation is helpful. Finally, researchers should read the patent court decision, *Armstrong v. Emerson Radio and Phonograph Corporation*, 179 F. Supp. 95, Southern District of New York, 1959. This document restates the argument that Armstrong invented FM radio alone, but the judge also reviews evidence on both sides at length, something Lessing never does.

To understand the history of FM radio broadcasting, one must know about the history of communications technology and radio broadcasting in general. Since the early twentieth century, scholars of radio have had at their disposal a number of good general histories of radio broadcasting. In addition to Lewis's *Empire of the Air*, these include Gleason L. Archer, *Big Business and Radio* (New York: American Historical Society, 1939); Erik Barnouw, *A Tower in Babel: A History of Broadcasting in the United States to 1933* (New York: Oxford University Press, 1966; repr., 1978); Christopher H. Sterling and John Michael Kittross, *Stay Tuned: A Concise History of American Broadcasting* (Mahwah, N.J.: Lawrence Erlbaum Associates, 2001); W. Rupert Maclaurin, *Invention and Innovation in the Radio Industry* (New York: Macmillan, 1949). A good history of radio from the perspective of audiences is Susan J. Douglas, *Listening In: Radio and the American Imagination, from Amos 'n' Andy and Edward R. Murrow to Wolfman Jack and Howard Stern* (New York: Times Books, 1999). See also Hugh G. J. Aitken, *Syntony and Spark: The Origins of Radio* (Princeton: Princeton University Press, 1976), and *The Continuous Wave: Technology and American Radio, 1900–1932* (Princeton: Princeton University Press, 1985); Susan J. Douglas, *Inventing American Broadcasting, 1899–1922* (Baltimore: Johns Hopkins University Press, 1987); Sungook Hong, *Wireless: From Marconi's Black Box to the Audion* (Cambridge, Mass.: MIT Press, 2001). Among the best histories of radio and telephone technology are M. D. Fagen, *History of Engineering and Science in the Bell System: The Early Years (1875–1925)* (New York: Bell Laboratories, 1984), and S. Millman, *History of Engineering and Science in the Bell System: Communications Sciences (1925–1980)* (New York: Bell Laboratories, 1984). Another useful source is G. G. Blake, *History of Radio Telegraphy and Telephony* (London: Chapman & Hall, 1928).

Secondary sources about the regulation of radio broadcasting during the 1920s include Susan Smulyan, *Selling Radio: The Commercialization of American Broadcasting, 1920–1934* (Washington, D.C.: Smithsonian Institution Press, 1994); Slotten, *Radio and Television Regulation*; Marvin R. Bensman, *The Beginning of Broadcast Regulation in the Twentieth Century* (Jefferson, N.C.: McFarland, 2000); and Robert W. McChesney, *Telecommunications, Mass Media, and Democracy: The Battle for the Control of U.S. Broadcasting, 1928–1935* (New York: Oxford Press, 1993). For a history of clear-channel broadcasting, see James C. Foust, *Big Voices of the Air: The Battle over Clear Channel Radio* (Ames: Iowa State University Press, 2000).

Other secondary sources are available at the IEEE History Center, which provides transcripts of interviews with communications pioneers. Numerous Web sites replicate primary sources dating back to the nineteenth century. Especially helpful is

Thomas H. White's "United States Early Radio History" at http://earlyradiohistory.us/, although many of this site's articles lack volume numbers required to make complete scholarly citations. Susan Douglas's *Inventing American Broadcasting* examines gender and the origins of amateur radio. For a study of the growth of American corporate research, see David F. Noble, *America by Design: Science, Technology, and the Rise of Corporate Capitalism* (New York: Oxford University Press, 1977).

INDEX